DIMENSIONS OF TIME

DIMENSIONS OF TIME

The Structures of the Time of Humans,
of the World, and of God

WOLFGANG ACHTNER
STEFAN KUNZ
and
THOMAS WALTER

Translated by

Arthur H. Williams, Jr.

WILLIAM B. EERDMANS PUBLISHING COMPANY
GRAND RAPIDS, MICHIGAN / CAMBRIDGE, U.K.

Originally published in German as *Dimensionen der Zeit,*
© 1998 by Wissenschaftliche Buchgesellschaft, Darmstadt.

This edition © 2002 Wm. B. Eerdmans Publishing Co.

Wm. B. Eerdmans Publishing Co.
255 Jefferson Ave. S.E., Grand Rapids, Michigan 49503 /
P.O. Box 163, Cambridge CB3 9PU U.K.

Printed in the United States of America

07 06 05 04 03 02   7 6 5 4 3 2 1

Library of Congress Cataloging-in-Publication Data

Achtner, Wolfgang, 1957-
[Dimensionen der Zeit. English]
Dimensions of time: the structures of the time of humans, of the world,
and of God / Wolfgang Achtner, Stefan Kunz, and Thomas Walter;
translated by Arthur H. Williams, Jr.
p.   cm.
Includes bibliographical references and indexes.
ISBN 0-8028-4998-9 (pbk. : alk. paper)
1. Time. 2. Time — Religious aspects — Christianity.
I. Kunz, Stefan, 1956- II. Walter, Thomas, 1966- III. Title.
BD638.A2513   2002
115 — dc21

2002024466

www.eerdmans.com

CONTENTS

Contents

ACKNOWLEDGMENTS

This book arose in the milieu of a study group called "Dialogue of Physicists and Theologians," to which the authors belong. This study group was primarily supported by ministers of the Protestant Church in Hessen and Nassau (EKHN), and by physicists primarily from the Laboratory for Heavy Ion Research (GSI) in Darmstadt. After a year-long program on the theme "Time in Physics," the authors independently continued to follow up on this topic for five years. The new concept of the tri-polar system of time arose from this process of discussion, and it is the basis of this book.

Our thanks go to the study group for its stimulating discussions, and to the two institutions of GSI and EKHN for their support of the group and their concern for interdisciplinary work.

For stimulating ideas and help with individual complexes of themes, we would like to thank the following persons: Egypt: Dr. Achim Müller, Worms; Mesopotamia and Egypt: Dr. Sergej Stadnikow, Tallin, Estonia; and music: Dr. Beate Regina Suchla, Göttingen.

Our thanks go to Dr. Arthur H. Williams, Jr. (Richmond, Virginia), especially for his deep commitment as translator, and also to Prof. Dietrich Ritschl (Basel, Switzerland) for his advice on the translation. Finally, we very much thank the Templeton Foundation for its generous underwriting of the expense of the translation.

GETTING IN THE MOOD

What's happening with time? There is a more or less conscious and general feeling that life and time are out of sync. On the one hand, until recently life was defined by the slogan "time is money," and life was very busy and hectic, with action-oriented managers as the prototype of success. They all had no time. On the other hand, in their ample leisure time modern human beings regress through stimulants of every type, and they are intensely attracted to slipping into depression by "killing time" to avoid effort, or by consumer lethargy. In a subliminal way a countermovement seems to be forming. The hectic alternation of hot and cold baths bring fatigue and lethargic sinking back into depression. Wisely, more and more people try to escape passivity in different ways.

- *Der Spiegel* reports in its edition of 3 October 1994 (N. 40, pp. 162-65) under the heading "Die Renaissance der Langsamkeit" ("The Renaissance of Slowness") that various groups have emphasized carrying out the slowness of life.
- The Klagenfurter "Society for the Delay of Time" champions the natural biological rhythms for the formation of social time. For instance, they have increased the time for covering 100 meters to half an hour.
- There is an increasing demand for courses on meditation for the regeneration of body and spirit and for consciousness training in houses of peace and monasteries. Seldom do businesses advertise without the slogan "here and now." Nostalgic review and continuous planning for

the future, however, rarely miss the occasional opting out of the lived present.

• Science has also looked into this topic. Histories of culture and critical considerations of culture (Gronemeyer 1993), scientific congresses (see, for example, Gumin and Meier 1989), and not least the "Society for the Study of Time" (Fraser 1972) ask about the altered comprehension of time and about tracking down the experience of time.

So, what's happening with time? Let us tarry a moment with the present symptoms. Goethe's Faust seems to anticipate this dialectic of hectic activism and the lethargic sagging of modern time. In a key scene, Faust concluded his pact with Mephistopheles (Faust I, 1699ff.). He was now under the direction of the devil's form of time, as he said:

If to the fleeting hour I say
"Remain, so fair thou art, remain!"
Then bind me with your fatal chain,
For I will perish in that day.
'Tis I for whom the bell shall toll,
Then you are free, your service done.
For me the clock shall fail, to ruin run,
And timeless night descend upon my soul.[1]

It becomes clear in this pact with the devil that Faust sacrificed the possible fullness — or also the feared emptiness? — of the immediate present for restless activity, which pointed to an uncertain future. From then on, under the cleverly restrained leadership of Mephistopheles, Faust rushed from one adventure of unfruitful scientific searching to the next. He always wanted to partake of hope and of the fullness and intensity of life. In ever-shorter ranges, his restless heart required of Mephistopheles ever-greater doses of intense experience. The hope for fullness of life was nevertheless weakened and clouded again and again. This hope was afflicted with the onset of the feeling of flatness and subliminal emptiness. However, once Faust got involved with the devil, he could no longer withdraw from the emptiness of this addiction to hyperactivity. He went from experience to experience until

---

1. Johann Wolfgang von Goethe, *Faust: Parts One and Two,* vol. 45 of *Great Books of the Western World,* ed. Mortimer J. Adler (Chicago: Encyclopaedia Britannica, Inc., 1990), p. 17. Translated by Philip Wayne (Penguin Classics, 1949, 1959).

it became clear to him that the hope for fullness of experience in this hyperactivity was a deception. It was arranged ingeniously by Mephistopheles and was laid out for the devil's own purposes. Mephistopheles also pursued a goal: he wanted Faust's soul. This goal became for Faust all the more his own. Faust slid ever more deeply into the emptiness of this diabolical hyperactivity, until in the end he burned out after many adventures. Burned out and entangled in guilt, Faust remarked that in the truest sense of the word he was taken in by a diabolical deceit.

Faust had hoped to find fulfillment in large doses of experience and excitement, but in the end he found only emptiness. In an ambiguous misjudgment of the real circumstances, Faust had sacrificed the experience of the present ("If to the fleeting hour I say 'Remain, so fair thou art, remain!'") to a future that was itself a constantly rushing sequence of experiences. The present melted away for him. His spirit was always busy with the hope of experiencing fullness in the future. Again and again that hope proved to be a mirage.

Certainly, many contemporaries will find themselves perplexed at this attitude to life. Consciously or unconsciously, it seems that modern, uprooted human beings on average have been following a life-style drawn from Faust. They are subject to the ambivalence of empty activism and exhausted lethargy. In ever-shorter ranges and ever-higher doses, in vain they seek fulfillment in the intoxication of public mass events, or they are reduced to the level of a dull rising and falling of excitement. Also, they miss the lived present in favor of an imagined future. Experience is everything for them. It is a society of experience! They look for attraction, and the search turns into addiction. The doses must be constantly increased, and the intervals between experiences must constantly be shortened. If experience does not happen, then at least their feelings must be correct. These humans are increasingly addicted to searching, and yet they lose experience more and more. They go to self-help, self-discovery, and self-awareness groups as they look for balance itself. As long as they chase after experiences, they cannot find themselves, and they cannot have any experience at all.

Experience presupposes that we are in the present with alert senses. First, the future arises in an organic manner from a perceived and formed present, and then it "is kept" in the following history, and not "overtaken." Thus, the matter of the experience of time is also the matter of attitude. Neither the rescuing of the future nor the clinging of humans is required, but rather a letting go in the present, so restlessness stops and rest is found.

In anticipating our time, Goethe seems to have felt the same way. After drawing the unlucky lot of restlessness, he sought luck to draw rest from the moment. So, he expressed his changes of mind in one of his most beautiful poems, which is unsurpassed:

A Likeness

Over all summits
There is rest,
In all spinning tops
You sense
Hardly any breath;
The birds are silent in the forest.
Only wait, soon
You also rest.

The railroad train has an endpoint; the present time stops for the night. In another poem from Goethe, a similar concern was addressed about the future intervening in the present. It was set to music in Joseph Haydn's "Canon":

If you always want to wander,
Look at the good that lies so near,
Learn only to seize luck,
Because luck is always there
. . . always, always there.

In both texts, a turn to the future evolves from the immediacy of the present. As an analogy, today we find an effort to practice the experience of the present in the increasingly numerous meditation centers. The present is the goal to be found. However, the lofty goal is to reach it without further ado. Thus, we first practice forms of self-perception, especially the natural biological rhythms. Again, in this area we learn that Goethe was a pioneer. Characteristically in the "West-östlichen Diwan" we read:

Book of the singer

Two sorts of mercies are in breaths.
The air pulls in, then releases itself;
That presses, this refreshes;
So wonderfully is life mixed.

4

Thank God, if he presses you,
And thank him, if he releases you again.

Goethe therefore anticipated not only the situation of the modern perception of time, but in his poems also sketched two solutions:

- Leaping into a mystic time of the present.
- Perceiving and getting involved in the natural rhythms of human nature.

Are these solutions tolerable also for us? Or is it worthwhile to follow our paradigms of the economics of time and the expectation of the future? Right at the beginning, we do not want to put forward alternatives that reach too far too quickly. Qualitative considerations, which we just used, can best serve to set the mood of the situation. In view of the *crisis of time,* such a difficult problem makes it necessary to keep our eyes open for as many options as possible. That's why a sound diagnostic and therapeutic base is required for the sake of a *theory-controlled perception* that satisfies the following criteria:

- It must make comprehensible the structure of bygone human experiences of time.
- It must be able to make sense of the variety of the present human experiences of time.
- It must be able to provide tolerable solutions for our crisis of time.

In order to attain such a theory-controlled perception, we need a *model* of human beings that covers many aspects of their lives and that at the same time is quite open to the experience of time. We also need to focus sharply on individual domains of knowledge.

The Tri-polar Structure of Time

DEVELOPMENT OF A
THEORETICAL APPROACH

Our chapter title signals that we see humans in a threefold structure of relationships, which we illustrate with the help of a diagram (fig. 1):

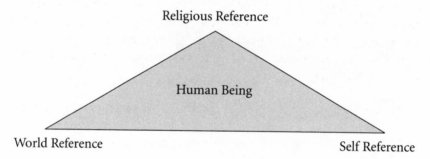

Figure 1

Of course, this model is very formal and vague at first, but it should prevent us from hastily reducing human life to just one of its aspects. Every model is a simplification and reduction of reality. However, a model can help us to find a perspective on a selected aspect, and such is the case with ours. It points out the interrelation of these three variables, and it will become clear in the course of our investigation how important this interrelation is. If we now apply this general model to our situation, it leads to the visualization in figure 2 (p. 7).

We call this model the *tri-polar structure of time,* and it will serve as a

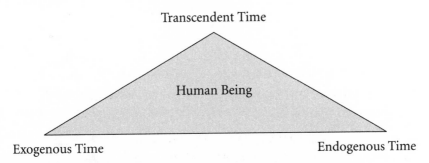

Figure 2

heuristic principle. We want consciously to leave the question open as to whether it is possible for one or two poles of this tri-polar tension to be reduced to one of the others. This matter must be decided on a case-by-case basis. However, it is crucial that we assume a relationship for the three poles. Such a conception is not generally accepted, especially since it is at the center of the philosophical discussion about the nature of time. Is time objective or subjective? In our conception, we assume that it is both subjective and objective, and that both poles are on top of one another. Of course, this conception must be proved at the empirical level. If, however, we accept this conceptual construction as a beginning, then we can ask all sorts of interesting questions:

- What would happen if there were a different accentuation of the three poles?
- Under what conditions is the tri-polar structure stable, and under what conditions is it unstable?
- What conditions lead to a fertile dynamics of the system, and what conditions lead to a neutral system?
- What are the consequences of synchronism, or asynchronism, for the whole tri-polar system?
- Which of the three poles stimulate the system, and which retard it?

With these questions in the back of our minds, we now want to describe the times of the three poles in detail.

ENDOGENOUS TIME

By endogenous time, we mean the forms of time that are accessible to human beings through immediate, inner experience. This time is applicable to the biological basis of all the experience of time, that is, the experience of all human beings together. Presumably this experience has not changed essentially in the course of the last millennia. We will especially look at elementary biological rhythms, such as inhaling, exhaling, the rhythm of day and night, the function of the thyroid gland in relying on this rhythm, and so forth.

Interestingly, however, it now appears to us that there is a basic biological continuity in the development of consciousness in human beings. That is, the course of the evolution of human consciousness is *overlaid with a conscious perception of time.* We now distinguish, in principle, three stages of that evolution of consciousness. These stages are coupled with a specific *apperception of time* and are first presented purely phenomenologically and rather vaguely. Later we will ask about their biological bases and will expound on them extensively through examples.

- *Mythic-cyclic time:* This form of the experience of time is especially found in the early, ancient cultures in the ancient Orient and in Egypt. Probably this stage of the evolution of consciousness is repeated in the childhood of modern human beings, before the maturing humans adapt themselves to the form of time that binds us socially.
- *Rational-linear time:* In this category the experience of time perceives the course of time primarily as an equally proceeding line. This is the form of the experience of time that is present in our society.
- *Mystic-holistic time:* Hereby we mean the specific form of the experience of timelessness.[1] This form of time is completely exempt from being in the present, as the mystics of all religions describe it.

These three forms of the perception of time correspond with each specific level of the evolution of consciousness. The tendency of the evolution

1. This timelessness has already been confirmed by neuroscience. The neuroscientists Andrew B. Newberg and Eugene G. d'Aquili claim that a person with mystical experience "loses all awareness of discrete limited being and of the passage of time" (d'Aquili and Newberg 1999, 110).

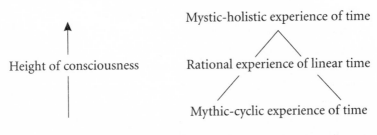

Figure 3

is marked by a development from the simple to the complex. In each case the higher level encloses the antecedent one as a special case. Figure 3 (p. 9) illustrates this tendency schematically.

EXOGENOUS TIME

By exogenous time, we understand all the forms of time in which human beings walk in relationship to the environment. We subdivide such exogenous times into natural time and social time.

By natural time, we understand the rhythms of nature in the widest sense, such as the rhythms of day and night, the seasons, the ebb and flood of tides, the cycles of the moon and planets, and so forth. These rhythms are cyclic and distinguish themselves by their relative stability. By social time, we mean the socially accepted idea of time and the formation of time. In each case, it will depend on the values of the concerned society, which values developed historically.

With reference to the framework of our examination, we have some interesting results. For example, the relationship of natural time and social time comes within exogenous time. Remarkably, it appears to us that socially accepted time obtained its orientation over the centuries by the life of one single human being in our Western society: Jesus of Nazareth. The result of this alignment is the church year, in which the essential stations of the human formation of time, such as birth, everyday life, joy and sorrow, and, finally, death, have found their place and meaning and orientation. Of course, this predetermined scheme found different interpretations and centers of attention again and again. This formation of time is still socially accepted in modern times, although it plays hardly any role in the actual formation of time. Our socially accepted form of time is that of the dynamic, linear faith in progress. However, it is clear that many other forms of time can be considered as socially accepted.

TRANSCENDENT TIME

By transcendent time, we understand all those phenomena that can be interpreted in the widest sense as a religious experience of time. Especially, we consider the experience of mystic, prophetic, and epiphanic times since they still are the most accessible for a description on the basis of experience.

The mystic experience of time, with which we have already dealt under endogenous time, returns once again. It is unclear whether this experience is purely self-awareness or whether cosmic[2] or transcendent factors are also involved.

We understand the prophetic experience of time as attested in the announcement of future acts of God by the biblical prophets.[3] Transcendent and anthropological factors mingle in this experience. Often it is not even clear to the prophet which factor goes back to transcendent intuition and which to some human imagining. Also, for this reason, prophetic statements about the future were not always applicable in the Bible. This prophetic experience of time is a mixed form, which is between endogenous and transcendent time.

By epiphanic time, we mean the real time in the Bible. The divine fullness of time and the divine possibilities of time were breaking into human

2. In numerous witnesses of mystic experience, again and again the experience of cosmic unity and the overcoming of the split between subject and object is emphasized (see, e.g., Capra 1980 and Kapleau 1989, 266ff.). Occasionally also a gaining of knowledge is claimed for this experience, and Eastern mysticism and Western physics converge on this conception of unity, as well as the experience of unity. If this interpretation should apply to the mystic experience of cosmic unity, then we would have to interpret this experience in our terminology as synchronization on the mystic level between endogenous time and exogenous natural time. However, there are also attempts to denote this cosmic character as pure self-awareness without any objective value (see, e.g., Schüttler 1974, 39ff., 114ff.). In this case, mystic experience would have no objective value apart from self-knowledge.

3. We strictly differentiate the prophetic experience of time from those outside the Bible. Mantic oracles and divinations are prophetic, but they have nothing to do with biblical prophecy, and they were also expressly fought by biblical prophets (e.g., Isa. 44:25; Mic. 3:7).

experience. They came to expression in the Old Testament: for example, Moses ("I will be what I will be," Exod. 3:14 NRSV, using alternative translation), and Jesus Christ in the New Testament ("the kingdom of God has come near," Mark 1:15 NRSV). By epiphanic time we understand a transcendent time that is clearly recognized in its effect on and in its relationship to humans in history. This experiential, transcendent time will be addressed from the human point of view in the chapter "The Time of Humans," and also again in a narrower sense from a transcendent view in the chapter "The Time of God." Naturally, since meaningful statements are possible only with regard to the effects of transcendent time with respect to human beings, we restrain ourselves from speculations, such as viewing transcendent time independently of its external effects.

SYNCHRONISM AND ASYNCHRONISM IN THE TRI-POLAR STRUCTURE OF TIME

The conceptual instruments developed so far allow us to view ideal and typical combinations of the three poles of time from the aspect of a synchronous and an asynchronous connection. Which connections are possible and which have been realized historically? If we simply ignore the fascist and communist social systems, we will be able to assume that all three factors are involved with each historically realized form of time, even if there are different priorities. The effects of the absence of transcendent time in totalitarian systems need a separate examination.

Before we pursue the subject of the ideal and typical combinations in the tri-polar system, we must make sure that these phenomenologically gained structures of time also have a basis in the basic biological structure of human beings. To put it briefly, is there a biological explanation for our three basic structures of the human experience of time?

The Time of Humans

TIME FROM AN ANTHROPOLOGICAL VIEW

BIOLOGICAL BASES OF THE EXPERIENCE OF TIME

Biological Rhythms

If we turn now to the biological bases of human time experience, then we must exclude transcendent time from our tri-polar system. It must be reduced to endogenous and exogenous times. Finally, the latter is reduced to exogenous natural time and exogenous social time. We therefore arrive at a bipolar system. In this system, we look at the individual rhythms within endogenous and exogenous times, and then we ask about their synchronism, as well as their overlapping nature. The science that deals with the exploration of such biological rhythms is called chronobiology (see, e.g., Mletzko and Mletzko 1985, 11ff.).

Within exogenous time, we can observe a series of natural rhythms. Their periods fluctuate from some hours (ebb and flood of tides) to some tens of thousands of years (Ice Ages). However, we restrict ourselves here to the ones that most probably may overlay endogenous rhythms. The most obvious one is the daily rhythm of day and night. In coastal regions, the rhythm of the tides follows this daily period. The phases of the moon are subject to a monthly period. The seasons are subject to annual changes, and the activity of the sun goes through an eleven-year rhythm. A period has also been determined for the change in the direction of the earth's magnetic fields (Mletzko and Mletzko 1985, 20-24). Furthermore, we can take into consideration even longer periods, such as the small Ice Ages, which appear approximately every ten thousand years. However, we want to refrain here from these longer cosmic periods. The four most important rhythms are the rhythm of day and night, the rhythm of the tides, the phases of the moon, and the rhythm of the seasons. These rhythms are also called circadian, circatidan, circalunar, and circaannual (Aschoff and Wever 1981).

Within endogenous time, the human organism is subject to a whole series of cycles. The only ones named here are those whose period amounts to approximately one day. Such periods are therefore called circadian, from *circa* = approximately, and *dies* = day. The most important of these rhythms is the rhythm of sleeping and waking. There is certain plasticity to this rhythm as it applies to humans, especially where it involves the duration and the time of sleep. Also, the endurance of humans, measured as body temperature, has a daily period. With young people, it has a low point at night, between 3:00 a.m. and 5:00 a.m., that is, during the phase of deep sleep, and it has a high point at 6:00 p.m. The kidney shows a daily period for the removal of potassium, sodium, and calcium. The same applies to blood pressure, as well as to hormone production. For example, cortisol is produced by the adrenal gland and serves for the preparation of energy. The effectiveness of medications has been proven to depend on the time of day,[1] and such matters are being explored by the newly specialized discipline of chronopharmacology (Lemmer 1983).

In the psychological area, daily rhythms are also detectable. Attention and the ability to concentrate are subject to a daily period, just as are the speed of reaction and retention. The latter is best stored in long-term memory in the morning, and then in the afternoon (Coleman 1986). Most people have a peak for their accomplishments in the late morning, with another peak, not a high, in the afternoon. These endogenous rhythms are partly related to one other. For example, body temperature is coupled with cortisol production and the rhythm of sleeping and waking. At night, cortisol production sinks to its lowest level. Its production rises continuously for some hours before awaking, so that energy is available to the nearly awakened body for the demands of the day upon awakening.

Are these endogenous and exogenous rhythms connected with one another? Are the endogenous rhythms influenced by the exogenous ones, or even steered by them? Are these rhythms somehow coordinated with

---

1. Overview: *Die Zeit* 49, no. 29 (November 1996): 34.

one another? An obvious supposition is that the rhythm of sleeping and waking is steered by the rhythm of day and night. However, by isolation experiments in subterranean laboratories, chronobiological research showed that the rhythm of sleeping and waking does not depend on the period of day and night. Rather, human organisms have an autonomous, endogenous rhythm of sleeping and waking, which is independent. According to test subjects, it is about an hour longer (about twenty-five hours) and about an hour shorter (about twenty-three hours) compared to the twenty-four-hour rhythm of day and night. The twenty-four-hour rhythm of sleeping and waking is therefore the result of synchronization between endogenous and exogenous time. If the external, exogenous timer is discontinued, then endogenous time runs freely, in accordance with its own rhythm of twenty-three hours, as well as twenty-five hours. How and when this synchronization occurred is unresolved up to now.

For ages, it has been assumed that there is a dependence of the endogenous rhythms of humans on the exogenous period of the lunar phases. The female menstrual cycles have been assumed to be influenced especially by the phases of the moon. However, such investigations are to be viewed with great caution (Rubber 1948). The same applies to the period of the reversal of the earth's magnetic fields as an influence on the human rhythms of sleeping and waking. It has yet to be determined what influence geomagnetism has on human organs (Mletzko and Mletzko 1985, 43-44). Finally, the influence of the eleven-year rhythm of solar activity (sun spots) on human symptoms, such as tuberculosis, heart attacks, and so forth, is being investigated through experimentation (Mletzko and Mletzko 1985, 27).

Endogenous Rhythms and Psychic Illnesses

Of all the synchronisms and dependencies of endogenous time on exogenous time named here, the synchronism between the endogenous rhythm of sleeping and waking and the exogenous rhythm of day and night is decidedly the most important. This importance shows up when a delicate desynchronization occurs between the two rhythms. Such a desynchronization can be found when exogenous social time undertakes another beat of time.

This situation is especially the case with shift work. Here it turns out that the endogenous rhythms are not modifiable at will: shift workers have a significantly lower life expectancy than that of the remaining population. Also, the isolation experiments already mentioned represent an intervention into the coordination between endogenous and exogenous time. Not least, therefore, experimenters discovered the freewheeling beat of time in twenty-three and twenty-five hours in the endogenous rhythm of sleeping and waking. With some of the test subjects, however, this intervention into the balanced system of endogenous and exogenous time had another surprising consequence. It turns out that the coupling of two further endogenous rhythms had broken down. We have already seen that the rhythm of activity, measured by body temperature, is coupled with the rhythm of sleeping and waking. As expected, activity is lowest at night and highest in the daytime. Observations of some test subjects for longer periods of time, however, have shown that this coupling breaks down (Meier-Koll 1995, 64ff.). That is, at night (at bedtime) body temperature rises unexpectedly.

DEPRESSION

This observation also throws new light on psychic illnesses in the area of cyclophrenia, for example, manic depressive illnesses. Meier-Koll (1995, 56ff.) argues that a desynchronism between body temperature and the rhythm of sleeping and waking can lead to manic depressive rhythms. If this supposition were correct, then attempts would have to be made to synchronize the rhythms. These attempts could take the form of changing the length of one of the rhythms for an improvement of the symptoms. In fact, attempts to treat depression through sleep deprivation have had a therapeutic effect. Subjectively, depressed persons experience their "pathology of time" as a reduction of speed. They feel inert in their experience of the flow of time. They feel cut off from the future, and they feel a weakness of incentive. Objectively, this falling out of synchronism expressed itself in a disturbed circadian rhythm of the metabolism of noradrenaline, which is an important neurotransmitter (Heimann 1992, 76-77). As we saw above, the coupling between body temperature and the rhythm of sleeping and waking depends also on the synchronism with the rhythm of day and night. Therefore, depression can be interpreted as a consequence of a disintegration of the bipolar endogenous-exogenous time system.

SCHIZOPHRENIA

The slowing of the subjective experience of time mentioned above is exhibited even more strongly by schizophrenic patients. By means of drugs, such as

15

psilocybin and LSD, schizophrenic psychoses can be artificially induced. The test subjects practically solidify into a kind of prison of time. Temporal before and after is no longer realized by the test subjects. The present appears to be infinitely lengthened, so that the river of time slows drastically. The entire temporal pattern collapses because of the regression of the subject. Beginning with schizophrenia, there was an investigation into the correlation with neurotransmitters. A connection was found between the distribution of certain neurotransmitters and schizophrenia. The neurotransmitter dopamine was discovered (Tölle 1991, 211) to trigger an elevated activity of the brain, and this activity is elevated in schizophrenic episodes. Therefore, such materials can be used therapeutically, because they muffle the activity of the dopamine system (neuroleptic) in schizophrenic illnesses.

Also, schizophrenics exhibit a more complicated electroencephalogram (EEG) than healthy persons, and with the help of chaos theory this phenomenon has been interpreted as an uncontrollably elevated activity of the brain (Stadler, Kruse, and Carmesin 1996, 323ff.). At the same time, schizophrenic illness is connected with a loss of reality and with a retreat of the libido, and there is an extreme loss of synchronism between endogenous and exogenous time. Schizophrenics are not capable of synchronizing their own time with the time of the world (Heimann 1992, 63ff.).

These two examples from psychopathology demonstrate how important the construction of complex forms of time, and rhythms of time, apparently are for psychological hygiene. At first, we are speaking of endogenous time in the sense of a coordination of coupled, endogenous time, as in closed-loop control systems. Then, we are speaking of endogenous time in the sense of a coupling of these endogenous systems with exogenous time, in the form of exogenous natural time and exogenous social time. However, how is this coupling of the endogenous rhythms managed with social and natural exogenous time?

Neurophysiological Coupling of
Endogenous Time with Exogenous Time

The human inner system of time can be subdivided into three connected subsystems: (1) the photoreceptor system, in the eyes and their coupling; (2) in a narrower sense, the endogenous timer system in the suprachiasmatic nucleus (SCN), and (3) efferent systems of SCN in the bodily systems (mentioned above), which are subject to a circadian rhythm. We turn to the connection of (1) and (2), that is, the coupling of endogenous and exogenous time.

For the past thirty years, scientists have explored the mechanism that causes the coupling between endogenous and exogenous time. More exactly said, this coupling of the internal time system becomes adjusted to the rhythm of the time of day, light and darkness. Two interconnected neurophysiological systems play a role in this adjustment. In the rhythm of light and darkness, the eyes perceive the intensity of light, and this intensity is supplied to a small area of closely neighboring neurons in the interbrain above the crossing of the optic nerves. This area is called the suprachiasmatic nucleus or the SCN. The mode of action of this coupling mechanism has not yet been investigated in detail, but we do know the basics of how it works.

The changes of light and darkness, which affect the photoreceptors in the eye, also affect the production of the hormone melatonin in the pineal gland. Melatonin apparently is responsible for the timing of sleep. The concentration of melatonin in humans was found to be five times higher in the nighttime, as compared with daytime. In addition, seasonally in the Northern Hemisphere the role of melatonin appears to cause depressive irritation, which indirectly proves that melatonin is what synchronizes endogenous and exogenous time. This illness is classified as seasonal affective disorder, abbreviated as SAD. It shows up primarily in autumn and winter, and it is dependent on the degree of latitude above the equator. With a certain percentage of the population, it takes the form of lack of drive, increased need for sleep, irritability, and depressive despondency. It turns out that people with these seasonal symptoms reflect a significantly elevated level of melatonin during the day. By means of a special phototherapy of at least 2,500 luxes per day, this irritation could be remedied within 50 percent of the patients.

We therefore see that endogenous and exogenous times are joined loosely together, not only with the system structure, but also with neurochemical messenger materials, such as melatonin. However, the actual endogenous rhythm can become independent of the changes of light and darkness in the SCN. In that case, it seems there is a constant temporal "rate of firing" of neurons in the endogenous clock within the SCN. The details of the formation of this rhythm are still unexplored. It is especially unclear whether this rhythm comes into existence through individual cells in the SCN, through an interplay of electrochemical oscillations in the SCN, or even in connection with other sharing in the brain, for example, with the hypothalamus (Gillette 1991, 125ff.). Also, it is unclear whether the afferent system (3) passes on the rhythmic impulses of the SCN from the circadian rhythms. To be sure, a whole series of electrochemical substances were examined for effects, but the exact mode of action is not yet certain (Illnerova 1991, 197ff.). Altogether, the coupling of the systems in the body with the SCN can be illustrated in figure 4:

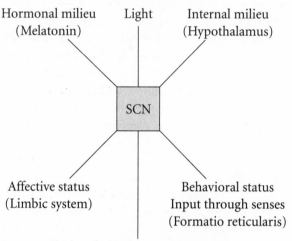

Hormonal milieu   Light   Internal milieu
(Melatonin)   (Hypothalamus)

SCN

Affective status   Behavioral status
(Limbic system)   Input through senses
                           (Formatio reticularis)

Hypothalamic homologous regulation
(Behavior of eating and drinking,
cycle of sleeping and waking,
cycle of temperature)

Figure 4

Time and Consciousness

Most of the temporal rhythms discussed to this point are not accessible for human experience. Nevertheless, humans do experience time, be it time as fleeting, or as tenaciously flowing, or as intensively rich in experience. In contrast to this conscious experience of time, however, the mainly unconscious processes, which are controlled by the SCN, run cyclic processes in human bodies, which have no organ that is identifiable with time. However, temporal experience is apparently closely connected with the ability of human beings to have consciousness. We must therefore make use of this relationship of time and consciousness.

This relationship of the experience of time and consciousness is recognized when changes of consciousness entail changes in the experience of time. The relationship is especially clear when consciousness is extinguished in sleep. Then the consciousness of time stops. During reveries, however, consciousness and awareness of time are present in very reduced form. With a diseased subconscious, as with schizophrenia, genuine depressions, or psychoses, there is, as has already been shown, a change from the normal experience of time. Finally, with drugs changes can be induced in the subconscious and in the experience of time. Especially by means of hallucinogens like mescaline, psilocybin, and LSD, there is an experience of extreme reduction in the speed of time (Heimann 1992, 66ff.; Huxley 1986).

Since consciousness is coupled with the maturing of the psychic system, it is not surprising that with children, whose consciousness still is in the process of the development, the consciousness of time matures gradually, and that with toddlers it does not exist at all (Piaget 1955). Vice versa, the experience of time itself changed with the evolution of consciousness. Consciousness is connected with the identity of "I" and the strength of "I," so that some authors speak also of an *I-time* (Heimann 1992, 65).

We are not surprised, therefore, that a change in the experience of time arose from the great thrust of subjectivization and individualization of the Enlightenment. Immanuel Kant (1724-1804) especially connected the experience of time with the activity of the "I" and with consciousness, when he in §2 of his "Transcendental Aesthetics" in *The Critique of Pure Reason* wrote concerning time (Kant 1990, B 49): "Time is nothing other than the form of the inner sense, i.e., the view of our selves and our inner condition. Because time cannot be a determiner of outer appearances."

This coupling of time with consciousness, and its anchoring in subjectivity, made an impression with the beginning of neo-Kantianism from 1850, and it shaped the nineteenth century. Neo-Kantianism hindered work on sensory physiology and an atomistic sensualism in Germany, while materialism flourished. On the basis of physiology, scientific psychology existed between 1850 and 1900, before Sigmund Freud's (1856-1939) discovery of the unconscious, thanks to the preeminent status of consciousness in Kant's psychology. This continuation of the tradition of Kant in psychology and physiology applied in particular to the founder of experimental psychology and physiology in the nineteenth century, Johannes Müller (1801-58). He believed he had proved the famous *law of the specific energy of the senses* by means of Kant's *a priori* principles.

Distinguished students of his, such as Ludwig Ferdinand von Helmholtz (1821-94), Robert Virchow (1821-1902), and Emil Du Bois-Reymond (1818-96), took up and continued his experimental impulses. The psychology of consciousness attained a climax and conclusion in the nineteenth century through Wilhelm Wundt (1832-1920), a short-term student of Johannes Müller. These physiologically oriented psychologists of consciousness brought no progress in understanding the connection between consciousness and time, as they thought they could identify the sensation of time with the duration of the sensation that accompanied it.

The philosopher Franz Brentano (1838-1917) provided an important expansion of this concept of consciousness. Familiar with the scholastic psychology bearing the imprint of Aristotle, as a former priest he recognized the *intentionality* of consciousness, that is, consciousness is never removed from its object. Consciousness is always consciousness *of* something. In this respect, consciousness always includes in itself a moment of will. In this way, Brentano established a bridge from bodily consciousness to objects that transcend consciousness. Brentano wanted to explain the formation of the consciousness of time through the effects of the rediscovery of the intentionality of consciousness. In contrast to his predecessors investigating the psychology of consciousness, Brentano did not identify the elements of the consciousness of time, which elements are the triggering stimuli. Rather he looked at intentionality, which has an effect on the imagination, and this he named *original associations*. From the stimuli of the accompanied sensations, the imagination causes the construction of the idea of time.

Brentano's student, Edmund Husserl (1859-1938), founder of phenomenology in philosophy, picked up this thought of the intentionality of consciousness, and he continued building on it. Above all, he raised the question as to how the objects of the world of experience are perceived, and how consciousness builds itself up and constitutes itself. This led him to the construction of his constitutional teachings. Temporality was included as an important element in the perception of the structure of consciousness.

The peculiarity of Husserl's opinion of the consciousness of time was to deny a loosely extendible, point by point consciousness of the present.

Rather, he thought that the consciousness of the present has from the start a certain extendedness. Therefore, *a priori* it covers a certain period of time, on which its intensity and vigor depends. He expounded this view in the example of hearing a melody. A melody could never be heard if its elements of time were built only from point to point in atoms of time. Rather, the perceptive consciousness still has present the sounds that have already gone past the present in certain respects in the hearing, and the coming sounds are already present in expectation. From this interlocking of memory and expectation, which Husserl called *retention* and *protention,* the continuum of consciousness builds the perception of time, which achieves attention. The intentionality of consciousness is shown in the intensity of the tension between the encroaching time in retention and protention. If this tension between retention and protention is large enough, then the consciousness of encroaching time can also supplement elements automatically into a whole. The impression of the integrity of a form can even suppress divergent details and lead to faulty perceptions. These thoughts were further worked out in the later Gestalt psychology and experimentally substantiated.

The connection between time and consciousness will now be illustrated. After World War II, neurophysiologists carried out numerous special investigations (Grüsser 1992, 79ff.) on the theme of time and consciousness. The work was performed partly by Grüsser, who continued from the attempts of von Helmholtz. There were investigations seeking to identify the smallest and largest portions of time perceptible by consciousness. Reaction time and estimation of the temporal duration of an event were measured. Also, neurophysiological theories were available to explain the *direction* of time neurophysiologically in the perception of consciousness. In this context, the exploration of the consciousness of the present by the neurophysiologist Ernst Pöppel (1985) of Munich is interesting. According to his research, in its perception of the present, consciousness holds events for only about three seconds. The present lasts three seconds, so to speak. Husserl's philosophical research was substantiated on the experimental level.

Important impulses have come from neurophysiological research on drugs and meditation, arising from the American movements of the 1960s to the 1980s (Haranjo and Ornstein 1988).

The Neurophysiological Basis of the
Systems of Consciousness of Time

In the exploration of the consciousness by modern neuroepistemology and research on the brain, intentionality finds in consciousness a generalization, which is renewed in a twofold meaning. In one meaning, consciousness can interact with itself and its subsystems (even referring itself to the consciousness of intentionality). In the other meaning, modern research on the brain has recognized the integration of consciousness into executions of acts in a system, in the sense of the somatosensory and the visual motor coupling of the eye-hand system. Thus, consciousness is *outside*, exogenous, and is bound into a circular process (fig. 5):

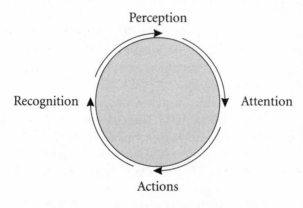

Perception

Recognition

Attention

Actions

Figure 5

Only in the execution of this circular process can a specific consciousness build itself and a consciousness of time. The consciousness built through it does not mirror the outside world in the sense of image theory, but it represents a particular structure of the incorporation of the outside world in action. It is clear that the respective consciousness of time depends on the particularity of the actions.

However, consciousness is also *inwards*, and the consciousness of time is incorporated in executions of internal neural actions. An unambiguous location of the consciousness of time in the brain is therefore not possible. Rather, the consciousness of time is a product of complexly interlocking executions of actions of sensory motor neurons. Therefore, consciousness

22

is distributed diffusely in the neocortex as a specific activity level, and it is coupled with external and internal sensory input and subsystems.

These subsystems are concerned with the *formatio reticularis*, the hippocampus, and the limbic system. The limbic system especially is one of the oldest systems, tribally and historically, because of its ability to activate attention, and it is closely connected with consciousness. By means of an EEG, a rhythmic oscillation of electric potential, the so-called α-rhythm with a frequency of 10 Hertz, can be derived from the neocortex. For the limbic system, it has been determined there is a rhythm of 5-7 Hertz, the so-called θ-rhythm. The following arguments speak in favor of this "localizing" of consciousness and attention in the neocortex, in connection with the subsystems named above (Grüsser 1992, 124ff.; Roth 1995, 192-225; Oeser and Seitelberger 1988, 137ff.):

1. The conditions of temporal disorientation always go along with a great disturbance of the EEG, which then shows a corresponding reduction of the speed of rhythmic paths.
2. In deep sleep, with perception of time turned off, the EEG accordingly shows a slow fluctuation of electric tension.
3. During the dream phases of sleep, if the consciousness of time exists rudimentarily, an increase occurs in the frequency of the EEG.
4. Infants and toddlers show very irregular EEGs. This situation corresponds to the fact that they still have not built a consciousness of time.

However, with children eight to ten years old, the periods of the EEG begin to stabilize between 8-9 Hertz, as the children build up their internal consciousness of time. This physiological situation corresponds to the fact that at first developing children form the perception of space, and then afterward they form the perception of time.

If it is true that the consciousness of time is integrated into executions of internal and external actions, and that it is also connected with a specific level of neural activity, then the consciousness of time must also be connected with a certain conversion of energy and of material in the brain. In fact, with the help of Positron Emission Tomography (PET), the glucose conversion in activated brain regions has been confirmed experimentally (Roth 1995, 200ff.). We thus come to a surprising realization: The consciousness of time is a function of the inner, activated condition of the neocortex. In principle, it is also connected with the limits imposed on the human brain by the metabolism of the conversion of energy, which makes changes in the cells of the brain. In fact, neuroscientists have already measured an increased blood flow in the prefrontal cortex (PFC) and a decreased blood flow in the posterior superior parietal lobe (PSPL) (d'Aquili, Newberg 1999, 119).

Let us survey this development of the consciousness of time with the background of our distinction between endogenous and exogenous time, which is yet to be established. We will see that an interlacing is possible between endogenous and exogenous time with increasing inner or outer action. Kant had determined purely the consciousness of time; Brentano added intentionality, which is an element of will; and Husserl recognized self-transcendence in the form of the tension of retention and protention. Modern neuroepistemology has widened consciousness through the interlacing of consciousness with bodily actions, and thereby it has widened the experience of time by the intentionality of action.

With the background of these results, the three structures of time, which at first were dealt with phenomenologically, appear reasonable: the mythic-cyclic, the rational-linear, and the mystic-holistic structures. These three structures of time reflect definite heights of neural activity with energy conversion in appropriate areas in the brain. As we saw, the level of activity of the consciousness can happen outwardly by going through the circular process of figure 5: perception-attention-action-recognition. Likewise, it can also be heightened and lowered inwardly by means of self-referential activity, such as meditation.

It seems that the lowest level of activity is connected with mythic-cyclic consciousness. This agrees with the view that the early mythic religions were extremely oriented to space. Sacred places and mountains, trees, and the epiphany of the divine in the space of the temple played a leading role. This mythic consciousness of time was simply not strong enough to dissolve the spell cast by space and to transcend itself into the future. Therefore, in the early mythic cultures the relationship of endogenous time to exogenous time was marked by the subjugation of endogenous time under exogenous natural time.

After the results that have been gained so far, a break away from this mythic spell seems possible in two ways. One way can be gained outwardly by continuing actions that transcend space. In the Western tradition, this way has been uniquely followed only by the Jewish religion, according to witnesses in the history of religion. The other way can be followed by the

24

constant removal of inner identification, as well as by withholding the functioning of consciousness. This is the way of consciousness training through meditation as it is practiced in different ways in the Eastern religions.

On the basis of our preliminary considerations of biology, it is now clear that bodily activity belongs to the practice of consciousness training, as in Yoga or Zen, in the form of breathing control, as well as of breathing observation. While the history of religion attests mystic experience[2] and there is research into modern meditation (Engel 1995),[3] the explanation[4] of a more exact physiological correlate for the change in the perception of time in meditation still awaits us.

The evolution of the perception of time by human beings has in every case been from the mythic-cyclic to the rational-linear to the mystic-holistic structure. Connected with this evolution was an increasing ability for self-transcendence, and also energetics and elevated metabolic activity of the body.[5] The experience of time is *one* aspect of a progressive evolution of human consciousness. Certainly, it is premature or speculative to want to identify in the structures of time a neurophysiological correlate for our three structures. One possibility shall nevertheless be brought into the discussion.

It could be that the mythic-cyclic time corresponds to the oldest (the tribal-historical consciousness) and the most distant structure of the brain, the limbic system, the *formatio reticularis,* and the SCN. At this stage, consciousness had not yet emancipated itself from the basic biological realities. The rational-linear consciousness of time would then correspond to the activity of the neocortex, while the mystic-holistic consciousness of time would be interpreted as a successful integration of the unconscious systems in the older tribal-historical regions of the brain in the neocortex with the SCN.

2. Cf., especially for the perception of time in mysticism, the sermons of Meister Eckhart: sermon 9, 11, 12, 35, 36, 40; also Achtner 1991, 126-45.

3. Also the elevated metabolism in the experience of mystic time could be proved experimentally on the basis of the β-waves in the EEG (cf. Engel 1995, 185, 194ff.).

4. Eventually chaos theory could be helpful for a conceptual understanding if it reaches its goal of formulating the process of meditation recursively and iteratively as a self-organizing system. Another goal is to identify the individual steps of meditation *(purgativa, illuminativa,* and *unitiva)* as attractors with corresponding parameters. It would be a great scientific breakthrough especially if the experience of "Absolute Unitary Being" (AUB) could be described by an attractor.

5. Already William James assumed that in mystic experiences the body is energetically activated: "mystical conditions may, therefore, render the soul more energetic" (James 1958, 318).

If we take as a starting point the neurophysiological interpretation of the three structures of the perception of time by humans, then there are apparently two possibilities for denoting the experience of timelessness by the mystic. On the one hand, it could be a return to older, distant consciousness, so to speak, to *sub*temporal regions of the brain. Then, the mystic experience of time would be a regression of consciousness.[6] On the other hand, it could be an entrance into younger regions of the brain, nearer to consciousness, so to speak, in *hyper*time. It would have been carried out in regions of the brain in the aforementioned integration of the neocortex and older regions. In this case one could say that mystic experiences of consciousness show progression.

On the basis of the increasing ability for self-transcendence and energetics we are inclined to the latter interpretation of the experience of mystic timelessness. As we saw, increasing self-transcendence can parallel the construction of the three structures of time. Also, the path inward is built by means of stimulation in passing externally through the described circular process. Both paths have been followed in the history of religion.

It appears that our phenomenological description of the three anthropological structures of time has gained a biological basis. We now come back, therefore, to our question about ideal and typical combinations of our tri-polar system. In our bipolar biological system, transcendent time returns as a third component. This view of time is important for the purely biological view of human beings. The possibility is not excluded that the structures of religious experience cannot be reduced to some *a priori* religiosity existing in human beings (for example, mysticism).

We want to investigate fairly extensively the beginning stage of the mythic-cyclic structure of time with its possible combinations, and we want to explore the conditions for the dynamic development of this stage. We now begin our discussion of the tri-polar system, our heuristic principle, with possibly the simplest case. How does our system behave if the structures of all three systems, endogenous, exogenous, and transcendent time, are shifted cyclically and synchronized? Actually, there was a culture in which these conditions were largely realized historically. It was the culture of ancient Egypt.

---

6. This opinion is represented in the older literature, partly explaining mystic phenomena psychopathologically, as, for example, Storch (1922), but also in the literature from the followers of H. J. Weitbrecht (e.g., Schüttler 1968).

THE THREE STAGES OF THE
HUMAN EXPERIENCE OF TIME

Mythic Experience of Time

The Experience of Mythic Stability in Egypt

The most extensive documents about the understanding of time by ancient humans are available to us from ancient Egypt. We first isolate exogenous time for purely methodical considerations. Then we view the endogenous aspect of tri-polar structure, and then we view the whole system.

a. Exogenous Time For exogenous time, there were four timers that were constitutive for natural time in Egypt. The path of the sun served for the division of the day and the year, the cycles of the moon set the dates for certain religious celebrations, the annual flood of the Nile marked the beginning of a new financial year, and at the same time the date was set when a new Pharaoh could be inducted into his office. In each case, the counting of time took its bearings by the number of years of his administration. Finally, the Egyptians also followed the so-called solar ascent of the star Sirius. This star is in the constellation of the dog, and the Egyptians called it Sothis. It becomes visible in the morning sky at the moment of the summer solstice, that is, with its solar ascent. For the Egyptians it was an event as fundamental as the summer solstice and the flooding of the Nile. The solar ascent of Sothis coincided in time with these events. Therefore, special religious rites were always observed during this period of time. Let us now look more closely at these four exogenous timers.

The daily path of the sun in the sky, which was always cloudless, was more than a way for Egyptians to determine the beginning and end of the day. The points of the beginning and end of the semicircular path of the sun from east to west marked for them at the same time the gates of life and death, the disintegration of life and the renewal of life for the sun. For Egyptians the West marked where the sun sinks in the desert, as the country of death, and the East, where the sun begins anew its path, as the country of life (Frankfort et al. 1954, 56). We see that space for Egyptians — just as time in myth and its rite representing it — instinctively lent a characteristic, qualitative impression. At specific spatial places, the threat of death and the regeneration of the sun were established for Egyptians. Of course,

27

Egyptians were very interested in the unscathed path of the sun, so that it would survive the realm of death in the west. They thought they could, and they must, support its path by cultic activity, which was celebrated daily. We will come back to that in the discussion of endogenous time.

Besides the sun, Egyptians got their bearings by the path of the moon. They especially used the moon for the dating of religious celebrations. In late Egyptian culture, the moon became very important for the deities associated with it: Thot and Chons. However, a problem arose with the combination of the sun and the moon as natural, exogenous timers because the periods set forth by these two heavenly bodies were irreconcilable. The course of the moon lasts 29 to 30 days. Twelve months would have an average of 354 days, yet the solar year has 365.25 days.[7]

Since for the ancient Egyptians the moon played a subordinate role as a timer, this lack of coordination between the lunar year and the solar year was not crucial. Instead, the Egyptians created for themselves an abstract solar year with twelve months of 30 days each. This artificial year was 360 days long. However, this solution had the mathematical disadvantage of not being completely compatible with the solar year with its 365.25 days. Therefore the Egyptians inserted 5 additional days, and they came up with a year of 365 days. It is interesting that the priests, who conceived this calendar purely from counting, clothed it in mythological language. This mythological transcription throws some interesting light on the structure of the Egyptians' consciousness, which in no way approached rational thinking. So, we read with Plutarch (Plutarch 1777, 7:385-509):

> One day Geb, the god of earth, and Nut, the goddess of heaven, entered into marriage. Then Ra, the god of the sun, cursed the goddess Nut, and he swore that their children would not be born in a month or a day of a year. The goddess Nut turned to the wise god Thot for help. Thot had won a game he had played with the goddess of the moon, and he had received one seventy-second part of the light of every day of the 360-day year. He put together five additional days from these parts, and he installed them after the months at the end of the year. So, the goddess Nut got five days, and she bore five children: Osiris, Horus, Seth, Isis and Nephthys, all of whom were also worshipped as gods.

7. The exact value for the sidereal year is 365.25636042 days; for the tropical year it is 365.24219879 days (cf. *dtv-Atlas zur Astronomie,* vol. 7: Auflage [1983]).

These five days were given to atone for Ra, the god of the sun. Since this time the solar year of 360 days was lengthened to 365 days, but because the goddess of the moon had gambled away five days, the lunar year was shortened from 360 to 355 days.

Besides the sun and the moon, Egyptians owed the security and the affluence of their lives to the reliability of the Nile's periodic flooding. This period proved to be useful.

The division into years got its orientation from these floods, which were crucial signposts. The Egyptians let the year begin with the day of the first full moon after the flooding of the Nile, which took place in the Julian calendar around the 19th July. They divided the year into three seasons (flood, sowing, and harvest) of four months each, and each month had thirty days. So, the abstract administrative calendar (mentioned above) was created. The first day of the flooding of the Nile on the 19th of July had a peculiarity, which was emphasized because of a coincidence with a constellation of the fixed stars. On the aforementioned day, Sirius became the brightest star in the sky on the morning before the sun was seen in the East, which was seen mythologically as the country of life. Sirius was seen for some minutes before the light of the rising sun outshone it. This solar ascent of Sothis was a dependable landmark for the determination of the flooding of the Nile. Simple farmers without knowledge of astronomy could recognize this star. That explains why the difference between their calendar dates and the days of the natural paths did not particularly disturb the Egyptians. The priesthood gave this astronomical event proper religious respect. On the wall of the temple of the goddess Hathor in Dendera was set forth in hieroglyphics on the ascent of Sothis: "The great Sothis shines in heaven, and the Nile goes over its banks."

The Egyptians strictly drew the periods of time from natural time, but it is quite remarkable how this periodic natural time worked its way into the area of social time. We can speak of a conversion of natural time to social time. For example, the coronation of a new Pharaoh often occurred on the day of the first flooding of the Nile, or rather of the first falling of the Nile (Whitrow 1991, 48). This influence of natural time on social time became especially clear in the cultic actions of the Pharaoh in supporting the path of the sun.

We will portray the actions more exactly below in order to show the correlation of exogenous and endogenous time. Since the Pharaoh was the cultic *pars pro toto* of the state, this conversion of natural time into social time had a direct effect on the stability of the commonwealth. The crucial signposts of social time were determined completely by natural time, so that exogenous time was structured completely by the period's natural time.

b. Endogenous Time   The ancient Egyptians did not have an abstract word for time. Rather, they always understood time concretely in the context of certain events or actions. They were not arranged as a temporal series of past, present, and future of abstract, linear time, as in the Indo-Germanic family of languages. Rather, they were arranged according to the feeling of isolation or nonisolation, as is usual with the Hamitic-Semitic languages, to which the Egyptian language belongs. This experience of time was similar to that of the Hebrews in being typically oriented to action. Action and its result of isolation, or nonisolation, was of interest to Egyptians, not its position in an abstract sequence of time. Abstract, conceptual thought could not yet emancipate itself from the immediate experience of the framework of this mythic worldview.

Accordingly, there was a word for time that described the ongoing aspect, *neheh,* and a word, *djet,* which showed the completed aspect of an action.[8] This understanding of time found its particular expression in the mythic solar theology before the Amarna period. We want to glance at this theology, in so far as it illuminates the understanding of time. For the Egyptians the sun was a god, who many times changed his form in the daytime and in the nighttime. In the morning, he was called Chepre, at midday Re and in the evening Atum. The time of day followed from the sun god's relationship shown in these different names for the same god. Chepre was called the "nascent" one, Re was "changing himself," Atum was called "the completion" (Assmann 1991, 47). Analogous to this continuous process of change of the sun, ancient Egyptians found the creation to

8. The meanings of the two designations for time are debated by Egyptologists. Among others, the following translations of *djet-neheh* are proposed: past–present; the far side of time–this side of time; linear time–cyclic time; spatial infinity–temporal infinity. We follow here the meaning of J. Assmann. In the larger context of the Hamitic-Semitic family of languages, he has distinguished between the aspect of the result, that is, the unending continuance that is final, and the virtual aspect, that is, the still present unending supply of time. However, this distinction is not related to cosmic but to inner processes (cf. Assmann 1991, 39-40; Assmann 1996, 25-40).

be subjugated to a periodic process of change, and with it — in this process of change — to be constantly endangered.

The dependability of the sequence of temporal events was not guaranteed for them through a unique act of creation. The periodic processes of nature, such as the rhythm of day and night and the course of the year, were particularly endangered, and so there was a necessity for continual renewal. Both aspects were expressed in the terms for time. For instance, the hour was called "the passing," the year, "the rejuvenating itself." Both names referred to cosmic processes.

The removal of isolation in the cycle was symbolized by the sun, and isolation by the sign for rock. They believed that the sun, as the giver of this natural rhythm, needed constant regeneration and renewal. The place of this regeneration was the rite staged in the cult with a strict attention paid to myth. The cult actualized the mythic primeval time and guaranteed the regeneration and continuity of time. In this sense, Jan Assmann wrote (Assmann 1991, 53): "Continuity is a dramatic concept of time, and it calls human beings to action. *Djet* and *neheh* must be continuously unified, in order to keep time going, and to guarantee continuity. This happens in the cults of the Egyptian solar shrines. Here, time is kept going and the path of the sun is completed in a cultic way in the form of a ritual calendar, which accompanies every passing hour with actions and recitations."

We already see, therefore, what a great meaning the cultic production of the myth had for the Egyptians. If it were missing, it would endanger the stability of time and the world would go back into chaos. This endangering was a theme in numerous texts (Assmann 1991, 54):

The earth is devastated
The sun doesn't rise
The moon hesitates, it exists no more
The ocean fluctuates, the country turns back
The river flows no more.

The Pharaoh had to recite the relevant texts every morning in the temple, in order to guarantee the daily continuity of the path of the sun, and also the continuity of time. Let us look at this process more closely.

The sun god Re played an important role not only for myth but also for actualizing the cult and for the understanding of time. As we saw, the sun god's identity was not at all secure within the daily routine. Every section of the day came with a special quality of time and peculiarity of the cultic drama. Nighttime was the place of the greatest danger for the sun god, and also for time. The sun god entered the underworld at night, and he needed special protection there against dangers that were lying in wait. If he got by these dangers, then he would be regenerated on the next day, and he could manage his tasks again.

The regeneration of time in the sun's battle against the powers of chaos each night in the underworld, and the twelve hours necessary to it, was portrayed in detail in the *Amduat,* and it was known in numerous books about the underworld. During this especially dangerous trip of the boat of the sun in the night ocean, the paralyzing gaze of the serpent Apophis was closed down. With help from Seth and Isis who cast a spell, Re bound and dismembered his adversary Apophis in the seventh hour. In this way he prevented the dissolution of the periodic temporality of the path of the sun into undifferentiated, chaotic impurity.

Of course, the victory in the underworld over the dragons at night did not last, and the battle had to be fought again every night. At the end of the nightly trip in the ocean, there were still two important events. As the two eyes of Re, the sun and moon were protected, and a snake with a falcon's head now lay on the boat of the sun god Re. Only at the twelfth hour was the actual regeneration and rejuvenation of the sun god established, and also temporality. Again, a meeting with a snake, named Ka, caused regeneration. As the old sun Re passed through the body of the Ka-snake in this last hour, it was rejuvenated to be the new day. This cult, myth, and rite of the *Amduat* was not accessible to ordinary Egyptians, for they were reserved for the Pharaoh alone. They were found in the kings' tombs of the new empire up to the time of the reformation of Echnaton. The special role of the Pharaoh for the myth becomes clear. Only from the end of the new empire did these representations become accessible to private people.

What does this myth and its cultic, ritualistic production mean for our question? At this point, we do not want to enter into the discussion about the meaning of the concept of myth, a discussion that is hardly clear. Here, the following interpretation is proposed: myth reflected a certain structure of the ancient human consciousness and a specific perception of time. This consciousness especially drew on its involvement in the periodic, rhythmic natural processes. Negatively, this formulation means that these early people did not yet have an "I" strong enough to emancipate itself from this cyclic involvement. Therefore, they needed to safeguard the redramatizing of the original creative event by the permanent cult in order to prevent dissolution into undifferentiated impurity, and to guarantee the continuity of time.

This view was expounded in more detail in the individual motifs of the *Amduat*. If we interpret the sun god Re as a symbol for consciousness, then his nightly ocean trip can be read as a cipher for the endangering of the conscious being and his temporal structure. Especially the twice-repeated meeting with Apophis and Ka, as snakes, can be interpreted as an endangering by regression. This regression would dissolve the consciousness and its temporal structure in an undifferentiated symbiosis with primeval time. The snake stood for the dissolving power of the unconscious, and it endangered the system of consciousness. This archaic-chthonic aspect must represent the oldest tribal-historical aspect of human beings.

Consciousness in the form of the sun god Re was dissolved and emancipated from these archaic roots. However, night by night it had to fight again to secure its viability in its ocean trip. There were two strategies for that occasion. In the *Amduat*, in the seventh hour consciousness sought at first to destroy the serpent, the archaic form of Apophis. It did not succeed, for that would mean an unfruitful separation from its own creative origins. In the twelfth hour, however, the regeneration was successful, so that consciousness went through the unconscious that had the form of Ka, a serpent, without being devoured by it. That is, the older tribal-historical part of the chthonic was integrated into the younger tribal part of the consciousness, or, rather, the consciousness was connected with its archaic roots in a creative way. The power of consciousness passed through the unconscious connected with it, and the union enabled a new start for consciousness. Therefore, the sun god's Re grew older during the day, until it was evening again, and in the night that followed he experienced in the threat of his dissolution also the chance for his regeneration. In conclusion, we can say that the mythic consciousness fluctuated cyclically back and forth between an endangering dissolution into a symbiotic undifferentiation and a weakly distinctive temporal permanence.

The cyclic understanding of time made it difficult to compose a generally binding calendar since time began anew with each new Pharaoh. While objective dating is easily possible with the concept of linear time, it became a problem for the cyclic concept of time. Thus, in Egypt no calendar was established in the manner of marking on a straight line of linear time. On the stele of kings there was only a relative dating according to the

33

number of years of the administration of the relevant Pharaoh. Even the diplomatic correspondence, for example, the Amarna letters, was undated (Aldred 1968, 216). Already in the time of Echnaton, the priesthood of the city Heliopolis had developed its local religion in the direction of solar henotheism. The gods Atum, Re, Chepre, and Harachte became only aspects of the highest sun god, Re. This tendency went so far that the priests joined the names of other local deities to that of the rising god's Re as seals of quality, so to speak (Aldred 1968, 249-50). However, it was left for Echnaton to reach the logical conclusion of solar henotheism. It led to solar monotheism, with Aton the solar disk as the single god.

Before we turn to the effects of the religious revolution of Echnaton on the perception of time, in conclusion we want to interpret this ancient mythic time in terms of our heuristic principle and to look at endogenous, exogenous, and transcendent time.

We saw that endogenous time had a cyclic character. Also, exogenous time went through a cycle in the form of the rhythm of day and night and — crucial for Egypt — the annual flooding of the Nile. Finally, transcendent time in the form of the sun god and his phases of regeneration was subordinated to this cycle. All three aspects of this tri-polar configuration of time had the same structure. In our religious-psychological view we interpret transcendent time as a moment of consciousness, that is, transcendent time led back to endogenous time. Strictly speaking, this structure of time had only two poles, and therefore was bipolar, and the two components of this dual structure were synchronized. *However, this synchronism did not occur automatically but, because of the weakness of endogenous consciousness, had to be renewed again and again by the cult.* Only through the cult was something created that was still more important than the synchronism of the dual structure of time: *stability.* Actually, the religiously secured structure of the state of ancient Egypt was distinguished by a millennium of stability. With the help of our heuristic principle, we are now able to formulate the situation as follows: the mythic temporal stage, which is distinguished by a synchronous, bipolar (tri-polar) structure of time, demonstrates great stability if it is secured by the cult.

The permanence of cultic *action* guaranteed the stability of the structure of time. In the mythic stage of time we can see that natural time and social time occur in exogenous time. The cult established the connection between the two. This realization can be formulated differently: god's absence, that is, an autonomous, transcendent factor of time, forced excessive cultic and ritual actions onto the mythic stage of time in order to uphold the stability of time. At this point we have recognized the Achilles' heel of this system. If the synchronizing factor of the cult failed for any reason, then a disintegration of the whole system into isolated partial systems was to be expected — then the whole system would be arranged again at a higher level. It was precisely from this kind of challenge that Echnaton's monotheistic revolution arose. It is exciting to follow Echnaton's actions in this difficult changeover. We now want to turn to his attempts at religious reform and subsequently to reinterpret it in terms of our heuristic principle.

c. Transcendent Time This stage of mythic consciousness was put in question by the religious revolution of Echnaton. In Echnaton's *Hymn to the Sun,* we find an impressive document of the first monotheism in human history. We want to look at individual motifs from the *Hymn to the Sun.*

In Echnaton's revolution, there was a historical attempt to overcome polytheism by means of monotheism:

> You alone are god, and no one is to be compared with you!
> You created the earth after your wish, quite alone.

Earlier among his predecessors, specifically under his father Amenophis III, there was a religious concentration on henotheism, with the sun god Re as uppermost god. Only under Echnaton was the sun god acknowledged as the single god. From being a relatively insignificant local deity, the god Aton advanced as the true sun god, and he claimed exclusivity in the new capital, Achetaton. This claim was expressed in an excessive manner, so that through Echnaton's countrywide efforts the name of Amun-Re, the highest deity until then, was left off of all significant cultic inscriptions. This claim of exclusivity by the new sun god, Aton, led to a prohibition of representing other gods in pictures. The whole rich symbolic world of the other gods was repressed. Only the sun disk remained, and it was given the name Aton *(jtn),* as a pale, abstract symbol of the new sun god. Understandably, the newly provided cult of the sun was no longer found in the mysterious semidarkness of the traditional temples, but rather in the light of day:

> All eyes are on you,
> whenever you are over the country as sun of the day.

There was no longer the necessity of a *cultic* drama with mythic connections. That was especially true of the cultic-dramatic representation of the nightly ocean voyage of this sun god. As the only god, it was not necessary for Aton to fight the dangerous serpent, Apophis, on his nightly underworld trip. Neither was it necessary for him to experience regeneration by means of the Ka-serpent through a symbiotic annexing of himself. Aton was the only one, and therefore he was the only one who lives. He did not need protection from battles or support by the cult for his existence. In place of the cultic drama, the sun chant appeared, and in it there was a sober, rational description of the nightly absence of the sun. The threat to his existence by the fight with the serpent, Apophis, was considerably softened from now on by the belief that the sun only rested at night.

As the only creatively living one, he transcended all dualism, which manifested itself for good or ill in the cultic and mythologically cemented polarities of day and night, male and female, chaos and order. In place of polarities threatening the creation, a dependable order of creation now arose:

You create the seasons, in which all your creatures grow, to allow
the winter, in order to chill them,
the (summer) scorching heat, so that they feel you.
You made the heaven distant,
in order to rise upon it and to look upon everything you created.

The *Book of the Dead*, or *Amduat*, vanished, and a serene *orientation to this world* took the place of a consciousness of a nightly threat. This orientation was especially expressed in a new artistic view of everyday reality, as well as in a playful perception of the reality of creation:

The whole country does its work.
All the livestock is content with its herbs,
Trees and herbs become green.
The birds flew up from their nests,
Their swinging praise your Ka.

It is crucial for our question that Aton was designated as the lord of time:

How effective are your plans, you eternal lord.

We come back to our exit question: how does this perception of time appear to be connected with this monotheistic revolution? At first, it seems that the path of the sun was taken out of the process of cyclic renewal according to the rhythm of day and night. Earlier it was said in the text of an hourly ritual about the path of the sun god (Assmann 1975, 50): "Handsome youth with wide step, the one who was born in the day, day by day, the one who is carried nightly by his mother in pregnancy, day by day. . . ." That is, here the sun god suffered time passively. Then, as reflected in a text from the Eighteenth Dynasty (1554-1305 B.C.), there was the active creation of time by the sun god (Assmann 1975, 51): "You drive over the heaven in order to create time and to enliven people and gods."

The battles with the serpent Apophis who threatened time were no longer necessary, and neither was the regenerating symbiosis with the serpent Ka. Cult and myth slipped away because the sun itself was creator of time. The immediate consequence of this process was that the daily cult of hourly prayers for the support of the path of the sun became superfluous because of the sun god's autonomy. Henceforth, the sun god managed his large, daily workload by his own strength.

> The day is short, the path is long:
> Miles of millions and hundreds of thousands.
> In a brief moment, you have completed it,
> when you sink, you have completed the hours. (Assmann 1975, 51)

The roles were now exchanged. Earlier, the sun god required the daily and nightly cult of humans for the maintenance of his cyclic path. Now he is the one who puts humans at his disposal, as the autonomous sovereign of life. As a measure and allotment of time for humans, endogenous time came from the sun god's cyclic path, which was made safe by the cult. The inner disconnecting of endogenous time from cyclic time led inevitably from a cyclic idea of time to a linear idea of time. Beginnings of the developing consciousness of linear time appeared, for example, by the time of the Amarna letters, which in themselves showed a developing consciousness of history. For the first time in Egypt's history, shortly before the Amarna letters, a new relation to the future appeared:

> I was clever for the future,
> I learned from yesterday.

which was written on a burial stele from the Eighteenth Dynasty. Now, humans were forced to place themselves in relation to stable linear time, whether they seized *possibilities* as free human beings, or whether they submitted themselves to a deterministic *fatalism*. Further study of the religious history of Egypt with regard to the god of fate, Jati, showed that the Egyptians were not in a position to rise up to the level of a freer form of time.

Summary

Now it is time to indicate the changes in the entire system in light of our heuristic principle. At first, the dual system was broadened by the autonomy of transcendent time to a tri-polar system. By virtue of its autonomous linearity, transcendent time was no longer synchronized with endogenous or exogenous time. On the contrary, precisely because it was autonomous, it turned into a disturbing factor for the synchronous, bipolar, mythic structure of time. With the integration of transcendent time as the synchronizing factor between endogenous and exogenous time, this disturbance then provided instability since the cult was no longer applicable because it was superfluous. The Achilles' heel of the mythic structure of time had been found. Here is the place to answer the question about the fate of the development of the tri-polar structure.

Historically, we see that transcendent time did not succeed with Echnaton; that is, the new linear time of Aton was not made broadly, socially binding, in order to transfer transcendent time into social time. He succeeded only in Amarna with a small section of the population, restricted in space and time, although he vigorously tried a general social enforcement throughout the whole country through draconian measures — carving out all traditional names of the sun god before Aton on the steles. At his behest, the new cult came to the whole country, and it introduced the new god with his new time, but after Echnaton's death it was abolished relatively quickly. The inertial strengths of the old system were apparently too strong. Nevertheless, the reform was not completely ineffective because the subsequent time did not turn back to the old system altogether. Rather, a further existence of the tri-polar system was found in the *byproducts of disintegration* in the post-Amarna time. The old gods and cults were introduced again, but the idea of linear time was partially preserved in the form of a deterministic fatalism. However, this concept of linear time was no longer bound to a concept of a transcendent god, for it was connected in a remarkable way with the moon god Chons, who was connected with the natural rhythms. On this basis we speak of this diversity as a by-product of disintegration.

As result of our observations, we can establish the following: Since the linearity of transcendent time was not on the same level as the cyclic time of endogenous and exogenous time, it could lead to a desynchronism of the tri-polar structure of time. This situation, however, led to an instability of the entire system, which could not be stabilized at a new level.

Our heuristic principle allows us to question whether other connections could lead to desynchronisms. At the stage of mythic time, the question can be answered easily. In exogenous natural time, there might be inconstant environmental conditions, so that no dependable rhythms and cycles were to be found. Such a possibility could happen when the environment itself exhibits noncyclic changes to a great extent. In fact, this variation happened in Mesopotamia.

Experiences of Mythic Instability in Mesopotamia

a. Exogenous Time   Mesopotamia is one of the oldest cultural landscapes on earth. In contrast to the very old Egyptian cultural landscape, Mesopotamia has experienced the political reign of eleven different peoples in the course of the millennia. It is to be expected that this great political instability in the country of the two rivers would have an effect on its experience regarding time. Not only were the political conditions in the country subject to continuous change and instability, but nature was as well.

In contrast to the stable rhythms of nature in Egypt, especially those of the Nile, unstable and untrustworthy circumstances prevailed in the steppes and the river landscapes of Mesopotamia. Such instability especially appeared in unpredictable floods and in high water levels of the Euphrates and Tigris Rivers at unfavorable times (Wendorff 1985, 19ff.; Whitrow 1991, 55f.), which made irrigation at the appropriate time difficult, and thus productive agriculture was fraught with hazards. Climatic fluctuations, torrential showers called storms of the steppe, also obeyed no calculable rhythm. The insecurity caused by these factors in the immediate surroundings exceptionally impeded humans from reaching a balanced, harmonious relationship with nature, in order to produce a synchronism between endogenous and exogenous time. Also, an extremely fortunate

39

harmonization between the earthly natural time and the cycles of the heavens, in the form of the synchronism of the ascent of Sirius with the first flooding of the Nile, was missing in Mesopotamia.

It was necessary to find an answer to this challenge of the unpredictable course of nature. The answer for the Mesopotamians was twofold. On the one hand, they tried to make unplannable nature plannable in an artificial way; that is, they began to impress on nature an artificial cultural landscape. Irrigation systems, stockpiles, the construction of an administrative system with experts of various types, the transmission of money, including a system of credit with horrendous interest rates, all these were signs of this attempt to overlay the unpredictable natural time with a plannable, artificially homogeneous time. Of course, this motif of planning makes sense only in the context of an idea of linear time. Actually the Mesopotamians found god supporting humans, even in their planning of their surroundings:[9]

> If you build, you serve your god,
> If you don't build, you don't serve your god.

The cultural advance could have led to the formation of an idea of linear time if two other factors had not been restraining forces. On the one hand, the Mesopotamians depended on their deities so very much that an autonomous, individual consciousness of time could hardly develop the corresponding responsibility for their own temporality (Frankfort et al. 1954, 205ff.).

On the other hand, the need for harmony was apparently so strong at the archaic-mythic stage of time that the Mesopotamians replaced the missing synchronism of the immediate environment with the synchronism of the overpowering harmony of the firmament. If it succeeded in building a bridge between the observable, dependable cycles of the stars and planets and human concerns, then the longed-for synchronism would be produced between endogenous and exogenous time. The Mesopotamians found this bridge in astrology.

While it may be abstruse for the enlightened, rational spirit, this connection was not at all beside the point for mythic thinking. On the contrary, a basic principle of mythological logic (the universally applicable

9. *Revue d'Assyriologie* XVII: 122, iii and iv, 5-8.

principle of the *pars pro toto*) provided the connection between heavenly movement and human action. This connection between the ordered harmony in heaven and the disordered circumstances on earth was made conspicuous in the buildings of the Babylonians. They built high towers to connect heaven and earth. As attested in the Bible, the famous tower at Babel belonged in the category of these towers, which were also erected for religious purposes.

Such a tower was called a ziggurat. It was used for astronomical observations, and much rational knowledge about the movements of planets and stars was collected through these observations. This knowledge also served in the preparation of calendars. As the basis of their calendar, the inhabitants of Mesopotamia discovered the cycles of the moon, and they created a lunar calendar. During the Sumerian epoch of Mesopotamia's history, each city had formed its own calendar system on the basis of the lunar cycles. Under the reign of Hammurabi (1792-1750 B.C.), many areas of public life were centralized. Hammurabi created a collection of legal rules that were binding for his whole dominion, and that were entered into history as the codex of Hammurabi.

Religious politics had an effect on this tendency toward centralization in that the urban god of Babylon, Marduk, became the god of the country. In the prologue of the text of Hammurabi, it is said that "the reign over the people" was handed over to Marduk. Under the Babylonian sovereign Nebuchadnezzar I (1123-1101 B.C.), the god Marduk managed the breakthrough into his position as ruler of the country, as the epic *Enuma Elish* related concerning the creation of the earth. In a complicated process of adapting Sumerian patterns up to the time of Nebuchadnezzar, a tendency towards centralization arose that had its beginning under Hammurabi; it also had an effect on the measuring of time. The lunar calendar of the city Ur was declared by Hammurabi to be binding for the entire country. However, this edict was not just an act of power politics or an administrative act of law; it is clear that ancient humans were not able to cross the borders of their mythic consciousness. The preparation of this lunar calendar was understood as work in the court of god. By further delegation as a human issue, it became the powerful knowledge of the priests.

In the Mesopotamian epic *Enuma Elish* — we will come back to it later — Marduk entrusted the moon God Shin with the measurement of time, and therefore with the task of producing a calendar (Frankfort et al. 1954, 202):

He called the moon to appear, he entrusted the night to him,
He turns him into a creature of the darkness to measure time,
And decorates him with a crown every month without fail.

The lunar year consisted of twelve months, half as 29 days, and half as 30 days, so that the first appearance of the crescent of the new moon in the evening counted as the beginning of the month. The months had following names: Nisannu, Airu, Simanu, Duuzu, Abu, Ululu, Tischritu, Arachsamna, Kislimu, Tebetu, Schabatu, and Addaru. Each month, however, was divided into four weeks of seven days. The last day was a taboo and was reserved for expiation, in order to calm down the fury of the god. The year began with the month Nisannu. The lunar year consisted of $12 \times 29.5 = 354$ days. For astronomical reasons, in the course of time the lunar calendar showed more and more errors between the beginning of the month and the appearance of the lunar crescent (Seleschnikow 1981, 87ff.). Already in the eighteenth century B.C., months were added at irregular intervals to equalize the difference. Thus, we hear in a mandate from this time (Seleschnikow 1981, 97): "Because the year is not complete, the moon, which begins now, gets the name of the second Ululu, and in Babylon one shall not write the 25th of Tischrituone, but the second Ululu."

The modern way of counting hours of the day goes back to the Babylonians. It was introduced in 320 B.C. by the Babylonian astronomer Kidinnu at the sun temple in Sippar, a temple city north of Babylon (Kroll 1988, 65). There he also discovered the irregularity of the calendar in a nineteen-year cycle. In seven years of this cycle he inserted intercalary months, and they were in the third, sixth, eighth, eleventh, fourteenth, sixteenth, and nineteenth years.

However, this rational knowing was supplemented by a complicated system of correspondences, which were arranged between the clear heavenly circumstances and the unfathomable earthly entanglements. Simply put, it is astrology. The powerful caste of priestly astrologers worked to achieve intercession. Before making politically and socially important decisions, it was worthwhile to consult the stars. This led to taking earthly history, with its substantial insecurity and unpredictability, as an artificial analogy to the predictable cycles of the planets and stars, and as an analogy to the unpredictable and immediate surroundings as well. Stars, planets, and the moon were regarded as gods. The Babylonians arranged definite areas of human life for them. The moon god Sin and the sun god Shamash were responsible for the year; the goddess Ishtar (= Venus) was connected with human life. For the area of unexpected natural events, we find the following analogizing, which is handed down from the time of the King Ammizaduga (1581-1561 B.C.) (Gericke 1993, 17-18): "If in the month of Arahsamna on the 10th day, Venus has vanished in the East, remains invisible for 2 months and 6 days, and appears again in the month Tebetu on the 16th, then the harvest of the country will thrive."

In this context there is a very interesting example of the efforts to make analogies in the area of political history. From the temple city Sippar, cuneiform fragments, which have been dated approximately to the third century B.C., were deciphered in the year 1925 (Schnabel 1925), and they give us information about the abilities of the Babylonians in astronomy. One of these fragments predicts a conjunction of the planets Jupiter, Saturn, Venus, and Mars in the year 7 B.C. This rare conjunction took place as it had been precisely calculated by Babylonian astronomers 300 years previously. It only occurs every 258 years, and it was last observed in 1940-41 (Kroll 1988, 65).

Why is this conjunction of planets so interesting? The German astronomer Johannes Kepler suspected that the star of Bethelehem (Matt. 2:1-12) was formed by these three close conjunctions of Jupiter and Saturn in the year 7 B.C.! This conjunction could explain not only the brightness but also the unique appearance and the year of the appearance, 7 B.C. Johannes Kepler was persuaded by his own theory of planets to accept the year of 7 B.C. as the likely year of the birth of Christ.

The astrological interpretation that the Babylonian priestly astrologers gave to this conjunction of planets is also quite remarkable. As we noted, the Babylonians had developed a system of correspondences between earthly concerns and the movements of the stars and planets. According to this system of correspondences, the following interpretation was given: not only for the Babylonians but throughout all antiquity Jupiter was the sovereign star of the world. Jupiter was the royal star. Even the name given in the New Testament suggests this interpretation. Of course, in the New Testament it is found only neutrally as "star," but the Old Syrian translation of the original text uses the word *kaukeba* for star, and this usage corresponds to the Babylonian *kakkabu*. In late Babylonian times, this name was used for the planet Jupiter. The Babylonians connected Saturn with the lands of *amurru*, that is, Syria, and therefore also Israel and Judea. The Greeks depended on the Babylonians *expressis verbis* for Saturn as the star of the Jews.

If we now combine the astronomic conjunction of these two planets with the astrological assessment, a stunning interpretation emerges: in the country Syria (= Israel, Judea) in the year 7 B.C., a royal sovereign of the world will be born! Thus, the visit of the three wise men from the East in Matthew 2:1-12 is more than an edifying story; there is a historic kernel of truth underlaying the event. In fact, there is more: many astrological-astronomical technical terms were used in the story itself. The three wise men were government astrologers from Babylon, and they were designated by the Greek word: *magoi* (= magicians). If this interpretation is true, then it is a significant witness to the *providentia dei*.

However, we go back to the understanding of time in Mesopotamia. In essence, we can say that, because of the superimposition of the teachings about astrological correspondences, the Babylonians could not have attained enough of cultural advances to understand linear time. The seed of the understanding of linear time was planted, but it was suffocated by astrology.

We saw that what happened in Mesopotamia because of astrology was similar to what happened in Egypt in the conversion of natural time to social time. Before significant political decisions were made, the king consulted his priestly astrologers, who were employed and paid by him. As an example, we take a prophecy of the king Shulgi of Ur, 2046-1998 B.C., which prophecy his astrologers put in his mouth (Beyerlin 1975, 144):

> The countries altogether are guessing in confusion.
> The man leaves his wife,
> And the wife leaves her husband.
> The mother will lock the door on her daughter.
> The property of Babel will go to Subartu and to the country of Assur.
> The king of Babel becomes the prince of Assur.
> The property of his palace, his property, are delivered to Assur.

Texts of this type are numerous and show that a consciousness of linear history could not be developed on the basis of astrological understanding. The contingency that is so characteristic of history was artificially eliminated.

The position of astronomy in relation to political situations became even clearer in an astrological text that dealt with the brightness of the planet Jupiter and that then interpreted this brightness as an indication of political

44

developments on earth (Jastrow 1905, 639-40): "If Jupiter is clouded over upon its appearance, then hostile (kings) keep peace. If Jupiter bears radiance, the king remains unscathed, well-being is found in the country, and the country is found to be blessed. If Jupiter is strong, then the king of Akkad will roam triumphantly. If Jupiter is strong, high tide and rain."

These approaches marking the larger political developments were only a part of social time, however. For everyday social life, public rites were just as important. We now come to the celebration of the new year, which was so essential for Mesopotamian society.

At the beginning of every year, the *celebration of the new year* was carried out by all the people, together with the priests and the king in Babylon, in the immediate surroundings of the ziggurat. Altogether it lasted twelve days. We can identify the formation of social time in three complexes of motifs (Wissowa and Kroll 1930, 1667-71):

- An important motif for this celebration consisted of the *urge to organize time*, which necessitated a constant cultural work, which was suspended for this respite. In place of planned and directed time, there was celebration and partying in *unstructured, orgiastic presentness.* This abrogation of the organizational force of time certainly presented a regression psychologically, in which the weak "I" was still unstructured in regard to a symbiosis with the masses. In so far as it was only temporary, regression was regulated, and it served the "I" and had a regenerative effect in this respect.

- A second motif became visible in the idea of *expiation.* Time was always connected with an accumulation of guilt. Guilt incriminated and locked out the future. It needed a creative mechanism to overcome it for the possibility of life. The Mesopotamians found this mechanism in a rite. In a ceremony the king was robbed of the regalia of his power, and the chief priest slapped him in the face. Then the king was led before the god's image, before whom as a penitent he requested mercy for himself and for the people. After his acquittal and with the absolution of the people, he got back the regalia of his power. On the evening of this day, a white bull was killed and burned. The connection with expiation was clear. In this cultic action the white bull — forerunner of the proverbial scapegoat — symbolically became the *guilt saddled with time.* In this ceremony, the guilt accumulated in time was removed, and so in this sense it brought regeneration.

- Finally, there was a third custom for the celebration of the new year. As in the Saturnalia of the ancient Rome, the social order was suspended for the duration of the twelve days. The farmhand became a lord and vice versa.

Generally, instead of the real king a type of royal jester was enthroned for one day, and then for one day there were allowed all manner of foolish freedoms — however, on the following day the price to pay for this foolish freedom would be one's life.

Altogether, we can say that the orgiastic presentness, the ceremony of expiation, and the annulment of social forces contributed to the regeneration of time. These three aspects of the regeneration of time were only the social exteriors of a deeply religious concept of the formation of time. We now want to examine this concept in more detail.

We said already that under the reign of Hammurabi (reigned ca. 1728-1686, or 1792-1751 B.C.), Babylon became the center of the kingdom, and the local city god Marduk — an early solar deity — advanced to become the highest god of the kingdom. This young, active, and storming god was employed in the celebration of the new year's day party through the religious processions and cleansing rites in the temple. In order to understand the meaning of these rites, we must first make clear the theogonic development of Marduk. In the litany of the cult, the *Enuma Elish,* which was recited at the New Year's day party,[10] he was dramatically presented as the king, as myth said it. Just as papal pronouncements today are named by their first two words, the title *Enuma Elish* is literally the beginning of the epic. A few verses are quoted (Mann 1985, 123ff.):

As the heavens were not named up there.
As the earth did not have any name below,
As Apsu himself, the original one, the originator of the gods,
Mummu Tiamat, who bore them all,
Their waters mixed into one . . .
There were the gods born of Apsu and Tiamat . . .

The first gods, Apsu and Tiamat, were seen as sexual. Apsu was male, Tiamat was female, and they were variously represented as monsters. At

10. The most careful reconstruction of the celebration of New Year's day is found in Frankfort 1955, 313ff.

the same time, they symbolized the original elements. Apsu was the freshwater ocean below the surface of the earth, and Tiamat was the abyss of saltwater. These two gods represented the first generation of gods.

Since the Mesopotamians had a distinct propensity for battles and intrigues of the gods, this tendency affected their epics that relate the genealogy of the gods. Apsu and Tiamat produced the second generation of gods, Lachmu and Lachamu. These latter gods were parents of a third generation from Anschar and Kischar, the totality of the heaven and the earth. The subsequent generation formed the triad of Anu, Enlil, and Ea, gods of heaven. Ea's son is our Marduk.

The fourth generation of gods was drawn from an extremely extravagant way of life. Rest and order and custom and civility were severely disturbed. The noise that came with this way of life was hardly bearable, and it induced the father of the gods, Apsu, against the will of Tiamat, to plan to destroy the young, ill-mannered gods. The plan was found out, and Ea killed Apsu and changed him back into water. Ea now moved into the freshwater palace of Apsu as her dwelling. In this palace, Ea bore her son Marduk. Meanwhile, Tiamat, out for revenge, created all sorts of monsters, and she formed an alliance with the remaining gods. She also lay with Kingu, from the second generation of gods, as a new lover. In the showdown, Marduk conquered Tiamat, and he created the heavens from her body and the earth's disk from the body parts of the monsters he cut up. The defeated gods allied with Tiamat chose Kingu as a scapegoat. His performance of expiation would unburden the allied gods who participated in the original wantonness. This performance of expiation was not enough to recompense both the original wantonness and the stability of the new world order of the heavens, and thus guarantee the earth. The stability of the new order remained endangered and therefore needed a permanent performance of expiation. Who would provide it? Myth said that Ea used the vitality of Kingu to institutionalize a permanent expiation: "From his blood, she (Ea) created mankind. She prescribed the service of the gods, in order to free them."[11]

The creation of people was a solution for a predicament in the drama of the gods. They were not created for their own sake, and they were not prepared as model creatures. In their veins flowed a mutinous and humiliated god's blood (Frankfort 1955, 202):

11. *Enuma Elish,* tablet IV.

I want to join veins and I want to create bones.
I want to create Lullu, "human being" is his name,
I want to form Lullu, the human being.
He is weighed down with the tribulation of the gods,
so that they breathe freely.

After the creation of human beings, Marduk completed his work. Then, he placed three hundred gods in their respective places in heaven, and three hundred on the earth, and he transferred the tasks for them. As thanks to Marduk for their victory, the mutinous gods built Marduk a city, Babylon, and they built there the temple Esagila, where the people had to do eternal expiation for the trespasses of the gods.

This expiation was carried out for the people year after year during the New Year's day celebration. It was performed by the king in the central rite in and about the main temple at Esagila. The first week of the twelve days of the New Year's day celebration consisted of the expiation of the whole people for the sins of the past year. In connection with it, the king had to humble himself before the high priest, and he received his absolution. Now a sacrifice followed. In a pit, the sacrifice of a white bull was carried out. Possibly this sacrifice symbolized Marduk, and it partially took on the rites of the old vegetation gods, once again the victorious gods of the Mesopotamian pantheon, such as Tammuz, Ianna, and Ishtar. Other interpretations assume this rite repeated the first fight against Kingu, in order to prevent the cosmos from sliding into chaos (Frankfort 1955, 221). In any case, the rite served to maintain the life and order of the cosmos. From the eighth to the tenth day, Marduk was represented as death and resurrection. The king led the statue of Marduk at the head of a procession of images of gods from the neighboring temples into a temple outside the city. Together with the statue of Marduk, the king then spent three days in the temple in a subterranean chamber. The victory over the power of death was displayed dramatically in the cult. It was probably understood as a reenactment of the battle of Marduk with Tiamat. During these days the idea prevailed that Marduk was caught in the body of Tiamat, and once again he had to secure the way to freedom.

This process can be interpreted psychologically. It can be seen as the confrontation of the consciousness of solar light with chthonic-musty consciousness. In this respect, the dramatic, cultic drama had a heroic as well as a regenerative aspect. It was heroic because the solar consciousness freed itself year after year from the power of the chthonic that pulled it downward. It was regenerative because the solar consciousness went back to its chthonic, creative roots and thereby prevented mental sterility. The analogies to the cultic process of the Egyptians catch our attention. The difference consisted only in that once the Egyptians under Echnaton dared to attempt to escape from this process in time, they awarded to Aton the sole divine dignity, and the process of regeneration was no longer considered to be necessary.

During these three critical days, the social forces were held in abeyance in the city, the cultural pressure for the formation of time was lifted, and *orgiastic presentness* prevailed. On the eleventh day, they were led back in triumphal procession into the city. The images of the gods were set up again in the temple on the roof of the ziggurat. The divine wedding followed, which was the rite of the accession to the throne and the regulation of fate.

What did this myth and its ritual dramatization mean for the formation of social time? How was the inward religious aspect portrayed in the outward social aspect? At first, we may suspect a distinctly tragic feeling about life in the myth. In the rite, human beings had to achieve an expiation for the sins of the gods — year after year. They were paying for the sins of others, and they were to serve the compulsion that is imposed on them in the rite. Time therefore had to appear to the Babylonians as a *collective prison,* and for its smooth working the prisoners were still responsible (although personally no guilt could be charged to them!). These prisoners therefore were committed to an annual repetition of expiation.

This eternal captivity in unending expiation was only moderated by the regenerative part of the cult. The deeply religious inner side of the three days of fighting over death and resurrection between Marduk and the king in the temple Esagila corresponded to the simultaneous annulment of the social bonds in the city. There was lust, game and dance, and annulment and exchange of social roles; in short, liberation from the burden of time occurred in orgiastic presentness. In the mechanism that formed social time — we would almost like to say it this way — the neurotic tension of time was made tolerable by the playful annulment of time. This eased the obsession with collective temporal pressure to a tolerable level by the relaxing of controls that generated social pressure. Only in this way could the individual survive. Now we come to endogenous time.

b. Endogenous Time   Endogenous time, that is, the individual's internal experience of time, can only be indirectly opened up. We have great reservations about doing it from the available phenomena. We ask again about the structure of the self-consciousness that hid behind all these expressions. In contrast to Egypt, the lives of the Mesopotamians present a very irregular picture in the sources available to us. Nevertheless, we suspect that in all the different appearances of exogenous time, there is a least common denominator, which suggests a cyclic, circular experience of time. Three reasons for this follow.

First, the Mesopotamians were forced to face the principal insecurity and unpredictability of everyday life in history by the analogy that presented astrology as a cycle and circle. Their self-consciousness was apparently not strong enough to break away from the power of cosmically experienced circular movements, nor could it automatically form time into unprotected openness. The second reason lies in the cultic regeneration of Marduk. In contrast, the Egyptian Aton had become completely autonomous, and he no longer went through the cultic process of death and renewal. However, the life of Marduk remained bound to the cult, even if it was only carried out once a year. In comparison with Egypt, Mesopotamia, we can conclude, had a weaker self-consciousness, which was not able to attain a concept of linear time and history. The third reason is found in the stability of the cosmos that provided an expiation to be performed with hopeless regularity. This compulsion to act also does not agree with an understanding of linear time with its corresponding strength of self. We therefore draw the conclusion that Mesopotamia's inhabitants continued to be imprisoned in a cyclical, circular experience of time.

The single motivation that could have led to the formation of a consciousness of linear time was the coercion to advance culturally in view of the difficult geographical and climatic conditions. This pressure to advance was subjected to the yoke of the astrology, and so it could not develop. At the beginning of Mesopotamia's culture there was a phase in which people sought to be free from the depressing load of temporality and transience. It was found in the so-called *Gilgamesh Epic,* which came from Mesopotamia's early Sumerian epoch. In it, one theme was the inevitability of human mortality, and an attempt was made to emancipate the endangered self by a heroic rebellion. With his friend, the natural man Enkidu, the mythic King Gilgamesh, who is probably a historic figure, heroically sought to survive all sorts of adventures. Before a dangerous meeting with the monster Chumbaba, Enkidu's courage dwindled, and Gilgamesh built him up (Schott 1980, 31; tablet II, lines 140-45):

> Gilgamesch opened his mouth and spoke to Enkidu.
> "Who, my friend, could climb to heaven?
> Only a god who sits eternally enthroned with Schamasch;
> The days of mankind, however, are numbered,
> The wind is in vain, whatever work it may do forever!
> Here, however, you are afraid of death!
> See what happens with the power of your heroism."

The hero ignored death because at this stage in the development of consciousness he could still comfort himself with the honor of posthumous fame. The knowledge of his own mortality did not bother Gilgamesh. Death was the fate of all, and one was able to moderate it by dying honorably. Only as his friend Enkidu died did Gilgamesh become conscious of the hardness of the fate of death. He then asked the question about eternal life and, with his shrewd mythical ancestor Utnapishtim, set out on a long and dangerous search to find this eternal life beyond the waters of death. Utnapishtim could not help because the unique circumstances that led to his immortality could not be repeated. However, Utnapishtim had some good advice for Gilgamesh. He knew of a rare plant on the seabed that granted immortality. With the ferryman Urshanabi, Gilgamesh made the dangerous trip. He dove down into the sea, and he found the herb and made his way back home. On the way, however, a snake stole and ate the newly gained and precious elixir of life. The snake won immortality (skin), and Gilgamesh and mankind remained mortal. Thus ends the *Gilgamesh Epic* with the tragic consciousness of the finiteness of temporal existence (Schott 1980, 97; tablet XI, lines 290-94): "In the respite, Gilgamesch sat down crying. The tears flowed down his face: 'Oh, advise me sailor Urschanabi! For whom, Urschanabi, did my poor self strive? For whom is my lifeblood shed? I did not create good for myself.'"

We find this tragedy of the experiencing of temporality in the later cosmogony and theogony of the *Enuma Elish*. We now want to turn to this poem.

c. Transcendent Time   The *Gilgamesh Epic* did not provide any grounds for transcendent time. This tragic poem was essentially oriented to the human experience of life. Mythic outlines containing themes of the origins of the cosmos and gods were located in a series of other epics, such as the myth of Enlil and Ninlil or the myth of Tilmun (Frankfort et al. 1954, 167ff.). For our purposes, the *Enuma Elish* epic was most revealing. This epic went through a rather complicated beginning, which we cannot discuss further here (Frankfort et al. 1954, 186ff.). We only note that the main form of the poem, concerning the god of Babylon, Marduk, was probably inserted later (in a text that already existed) in place of the god Enlil of Nippur.

Marduk distinguished himself with a powerful move to establish order. After the fortunate victory over Tiamat, he was the one who gave the cosmos a new order, and he apportioned the firmament under his vassal god. In this sense, one could say that Marduk was lord of the cosmic course of time. Marduk granted relatively autonomous local times to his subordinate gods: for example, Shin, the moon god, and Shamash, the sun god. As lord of time, Marduk assigned Shin the responsibility "to measure time," which is the responsibility to take over the natural calendar. Marduk, however, occupied this reign over the course of cosmic time only conditionally. The stability of the cosmos and of cosmic time depended unquestionably on the performance of expiation by the people. The continually perpetuating performance of expiation by humans limited Marduk's reign over cosmic time.

In a certain sense, Marduk was dependent on the cultic practice of mankind. This dependence appeared again, as Marduk was in need of the regenerating the cult. Also, nothing could be said about a reign of Marduk over historical time. Crucial historical events, possible human initiative, and even possible divine reign over time were prevented by a cosmic fatalism, which would bend cycles and prevent the formation of a free and open time. Since we have seen the development of human consciousness on the basis of our religious-psychological beginnings in the forms of the gods, we cannot speak in analogy with the religious history of Egypt of an autonomous transcendent time. We have now arrived at the point where we can examine the tri-polar structure of time.

Summary   As in Egypt, we saw that transcendent time was not applicable as an *autonomous formative power,* so that strictly speaking we can only analyze a bipolar structure of time. We turn again to our heuristic principle and analyze how endogenous time was restrained to three aspects of exogenous time.

Let us look at the immediate experience of the environment to see if we can establish an extreme asynchronism. With its few constants and cycles, the cyclic inner experience of time was left with no way of showing a parallelism with exogenous environmental time. On the basis of the fundamental synchronism in the culture of Egypt, we can postulate a distinct stability of the state. We actually found it to be attested empirically. In the area of Mesopotamia, however, we can postulate an extremely unstable political culture, and it also was found to be empirically attested. Yet we also saw that the Babylonians sought to compensate for this deficit by turning to calculable cycles of heaven. With the help of the correspondence between heaven and earth, astrology could generate an *artificial synchronism* between endogenous and exogenous environmental time. Because of the conversion of exogenous environmental time into the exogenous social time through astrology and rite, this artificial synchronism between endogenous time and exogenous time succeeded. Through these artificial synchronisms the state was able to establish temporary stability. The compulsory character of this artificial stability was revealed, however, in that it collapsed regularly each year in the Babylonian Saturnalia, with its unleashing of orgiastic presentness. In Egypt nothing was known of an unrestricted social turnabout. This attitude accorded well with the natural stability of the social structure because of the synchronism between endogenous and exogenous time.

With Mesopotamia, we met a culture in which the synchronism between the three poles of time was disturbed on the mythic level, mainly by unstable environmental factors. As a way to restore this disturbed synchronism, the Mesopotamians selected a regressive strategy in the form of astrology and a progressive one in the form of cultural work. Finally, it led to the threshold of the understanding of linear time, without their being able to cross it, however. The strength of the immanent pictures apparently was not great enough to bring them to the level of linear time.

In our system, a desynchronizing was possible between endogenous and exogenous time, even if the environment were unstable, but instability in the exogenous structure of time was evoked by the migration of people. Such structures appeared mainly with nomadic people. We examine next, therefore, the effects of the nomadic lifestyle of the Hebrews before their settlement in Canaan on our tri-polar system.

Departure from the Mythic Experience of Time by the Hebrews

Who were the Hebrews? Who were the Jews? Who were the children of Israel? To a much greater extent than with the Egyptians and the Mesopotamians, we are instructed very well by the detailed and interrelated coverage of the Bible regarding the history of this people. This fortunate situation with the source is still supported by the historical-critical examination of the Old Testament, which has only existed for the past two hundred years. Through it we move into the situation of not only being able to reconstruct with some probability the origins of the Bible as a literary-historical document, but also of being able to appreciate the peculiarity of the biblical religion in comparison with the surrounding people.

The first historical message about the Hebrews is found in the biblical report of Genesis 10:21, where a certain Eber is mentioned (maybe Eber meant "alien" or "foreigner"). The figure of Abraham, who was important in holy history, descended from Eber's son Peleg (Gen. 11:16-26). Abraham was also the first in the Bible to be called a Hebrew (Gen. 14:13). The early pages of the Bible tell about Abraham. He was a nomad who emigrated from the land and culture of Mesopotamia. Even if the results of scholarly study cast doubt about the historicity of these early stories about this patriarch, the assumption that the first Hebrews were nomads with small numbers of livestock, and that they were in search of country and food, is still not too far off the mark. After all, one of the oldest texts of the Old Testament, the so-called small historical credo, in Deuteronomy 26:5-9, marked their predecessors as "wandering Aramean[s]," that is, as a nomads.

This characterization of the ancient Hebrews as wandering around in small nomadic groups is supported also by reports from the ancient cultures, if the Hapiru named in numerous ancient Eastern texts can be identified with our Hebrews. Although scholarship still has not reached any final judgment on this point, the people called the Hapiru are mentioned in the Amarna letters of Egypt, in the Mari texts of Babylon, and by the Hittites of Asia Minor in the fourteenth century B.C. (Noth 1950, 38-39). With some likelihood, we can assume that before the time of their settlement in Canaan the Hebrews were a nomadic people. The nomadic life did not stop with the settlement. Above all, two historical events were buried deeply in the consciousness of this people: the slavery and the liberation in Egypt, and the Exile and liberation in Babylon. In the course of this varied

54

history, the Hebrews formed their consciousness of time. We now want to pursue this matter.

Primarily, there are three very different worlds we have in view. With its respective peculiarity, each is expected to have a specific influence on the respective structures of the experience of time. At first, we look at the original world of the nomadic life, then at the time of the settlement in the culture and country of Canaan, and finally at the two prominent turning points of slavery in Egypt and the Exile in Babylon. "The time of God" is discussed in a later chapter.

Nomadic people are distinguished by their search again and again for places to pasture their livestock in new environments. The monotony and loneliness of desert landscapes, rugged mountains, extreme differences of heights, oases with limited resources of food, sandstorms and sudden hurricane-like downpours, and finally moving along, as opposed to the secured world of the cultured lands — all of these form the background of nomadic people. Such a time of nomadic life for the people of Israel continued to have an impact for a very long time on them and their history; otherwise, stories of this nomadic life would not be have been handed down again and again over the centuries in the form of small historical credos (Deut. 26:5-9). Spatial flexibility and continuous changes form the constitutive elements of the experience of the nomad. How did these factors have an effect on the structure of *exogenous* time?

First, it is quite plausible that the seasonal and agricultural cycles, as we observed them in Egypt, fall by the wayside. In fact, one might expect that the Israelites would have experienced a special orientation at different places through the cosmic cycles, as the Mesopotamians did. If this orientation concerning a certain popular knowledge of stars and planets were lost, then it assumed a continuity of observation and theoretical penetration. Yet that is only possible with a certain stability of location, a situation that is in contrast with that of the nomads, but not with the Babylonians. On the basis of the specific environmental conditions of nomadic life, we can exclude cosmic, agricultural, and landscape cycles, which are always recurrent, as constant, imprisoning points for temporal orientations. Thus, in Israel's nomadic life natural time broke down as a timer within exogenous time.

However, one could not say that these early nomadic clans knew of no landmarks in nature for their management of time. Actually, there are some hints that the larger family associations of the patriarchs used the cy-

cles of the course of the moon to regulate time. Purely externally, this supposition is appropriate if only because the nomads on their walks in the
nightly landscapes of the desert were dependent on the light and the reli-
able return of the moon. New moon and full moon and the constant periods were suitable as external timers. In addition, there are further signs for
this assumption: at least one of these nomadic clans emigrated from ancient Mesopotamia. In the prehistory of the Bible (Genesis 1–11) two
Mesopotamian cities are named, Ur and Harran, from which the predecessors of the Hebrews emigrated. These two cities were dependent on the
moon god Sin. Therefore, it may reasonably be assumed that Israel many
times affirmed the practice of a moon cult. (For the appropriate sacrificial
and oracle information brought by the patriarchs from Mesopotamia, see,
e.g., Num. 10:10; 28:11-15; Amos 8:4-7; Ps. 80:4; Prov. 7:20; Isa. 1:13ff.;
Hos. 2:13.)

Another sign of the existence of an early orientation to the course of
the moon occurred in the name of one of the patriarchs. Laban means
white, and it was at the same time a name for the moon. The church father
Eusebius (ca. 260-340) reported that in Harran Abraham and his clan had
been worshippers of the lunar god named Sin (Robinson 1988, 69ff.).
However, these occurrences exhaust evidence about exogenous time in
early nomadic life. All further biblical sources were of later date, even if
they may go back to earlier stages in the formation of tradition. It may very
well be that Israel took over the cult of the moon much later, after the time
of the settlement in the land of the Canaanites, when the Israelites had
taken over fertility cults in which the lunar god played an important role as
a female fertility goddess. We will return to this topic later, but all in all, we
can say that the moon could not have been an impressive giver of external
time.

The connection of lunar festivals (cult) and the Sabbath is interesting
(2 Kings 4:23; Isa. 1:13; 66:23 Hos. 2:13; Amos 8:15). Since the postexilic
time, the celebration of the Sabbath has primarily had a social dimension.
It is motivated in terms of theological creation and holy history, and its
derivation from the tradition of the moon cult, that is, from cosmic exogenous time, has probably been rightly disputed (de Vaux 1960, 330ff.; Robinson 1988, 155). What now remains as an agent for the temporal orientation for this specific nomadic life, however, if we do not value very highly
the meaning of the moon?

The consciousness of the migrant nomad could not be fastened to constant factors of time. On the contrary, it always had to be prepared to establish itself in changing environmental conditions and for unforeseen events. However, these conditions of life indicated an extreme asynchronism between endogenous and exogenous time. Yet, it forced the nomad to an elevated attentiveness and thereby to an elevated achievement of consciousness. The asynchronism between endogenous and exogenous time can be understood as an enormous stimulant for the strengthening of consciousness. This achievement of consciousness cannot have been without effect on the perception of exogenous time and the structuring of endogenous time.

We now recognize this achievement of consciousness by the nomad in that he made a virtue out of the necessity for his temporal orientation. In this sense, the Hebrew nomad replaced the constant elements for temporal bearings, on which the people of Mesopotamian culture could fall back, and the people of Egypt too, with a succession of unique experiences. Thus, the unique, unrepeatable events, and the sequence of such events, became the constitutive element of temporal orientation. The ability to *experience* now had to be connected with a singular event if it was to be used for the temporal orientation. In addition, the event must always be presentable to the consciousness; that is, one must be able to *remember* the event and to utilize it for the future, if necessary. All these activities demanded and educated the consciousness in the sense of a linear permanence.

Our thesis therefore is: The extreme asynchronism between endogenous and exogenous time put great demands on the system of consciousness and required it, for the establishment of a temporal orientation, (a) to build a relatively stable permanence, (b) to keep present in memory events of the past, and (c) to establish itself always anew in the changing circumstances of the future.

Our thesis explains why memory and hope played such a central role in the Hebrew language and religion. Memories of events in the past and hope for positive events in the future placed *responsibility* on the one whose consciousness was being formed for this task, that is, on the one seeking an orientation in time. Moreover, these unique events took place primarily in a social context. *Social time* had authoritative weight within exogenous time. It created what is known as the historical dimension of Hebrew religion. What has been discussed on the basis of examples from the time of the nomads will now be discussed in the time of the so-called patriarchs.

a. Exogenous Time of the Nomads Because of numerous uncertainties and a narrow textual basis, we want to refrain from the thought that the lunar cult, and the moon as a natural giver of external timers, may have played a role with the early nomads. We prefer to concentrate on the *social dimension* of exogenous time. We have already briefly sketched the panorama of the life of nomads and easily identified the basic focus of their life: the constant struggle with the shortage of resources, that is, water and pasture. This concern determined the *area of action* of the nomads, so to speak. Besides the insecurity of nature, a second factor that brought difficulty to their existence was the shortage of the resources in the social area that constantly forced them into *compensatory action*. Under pressure from natural and social bottlenecks, nomads had to develop by their actions a strategy for survival again and again to secure their ecological niche. We have identified this *action* as the constitutive element of exogenous, social time. We opened up this perspective with the help of our heuristic principle. This result agrees very well with the basic consensus of many separate examinations of the early Hebrew understanding of time. The Hebrew language had no abstract word for time. Also, speculation about time or eternity was foreign to the Hebrew language.

Hebrew had three words for time, and in each case they illuminated specific aspects of time: *'et, 'olam,* and *roega*. Besides these words, the word *we'atta* was used, most importantly in prophetic texts (e.g., Isa. 43:1; 44:1; 49:5) for divine action in time. *'Et* (Jenni and Westermann 1979, 370ff.) meant a specific time that was characterized by a qualified event. Such an event could take place in the context of natural processes, as well as in the area of human action.[12] Chapter three of Ecclesiastes is famous for its view of the temporality of natural processes. In the New Revised Standard Version, we read:

> For everything there is a season, and a time for every matter under heaven: a time to be born, and a time to die; a time to plant, and a time to pluck up what is planted; a time to kill, and a time to heal; a time to break down, and a time to build up; a time to weep, and a time to laugh; a time to mourn, and a time to dance; a time to throw away stones, and a time to gather stones together; a time to embrace, and a time to refrain

---

12. A thorough exegesis of these words is given by James Barr (*Biblical Words for Time* [London: SCM, 1962; rev. ed., 1969]) on purely philological grounds.

from embracing; a time to seek, and a time to lose; a time to keep, and a time to throw away; a time to tear, and a time to sew; a time to keep silence, and a time to speak; a time to love, and a time to hate; a time for war, and a time for peace. What gain have the workers from their toil? I have seen the business that God has given to everyone to be busy with. He has made everything suitable for its time; moreover he has put a sense of past and future into their minds,[13] yet they cannot find out what God has done from the beginning to the end. (3:1-11)

This text from late Judaism (ca. 300-200 B.C.) already breathed the spirit of an elegant, skeptical resignation. A deterministic, even a predestined, understanding of time gleamed through here. Little was expected of human activity. In this respect, the text was actually an exception, or rather a product of the religious moldiness of the canon of the Old Testament. Psalm 1 was more characteristic of what was expected of human activity, for in it the connection of human action and time was clearly expressed, as well as the connection to social obligation and time. We read in the New Revised Standard Version:

1 Happy are those who do not follow the advice of the wicked, or take the path that sinners tread, or sit in the seat of scoffers;
2 but their delight is in the law of the LORD, and on his law they meditate day and night.
3 They are like trees planted by streams of water, which yield their fruit in its season, and their leaves do not wither. In all that they do, they prosper.
4 The wicked are not so, but are like the chaff that the wind drives away.
5 Therefore the wicked will not stand in the judgment, nor sinners in the congregation of the righteous;
6 for the LORD watches over the way of the righteous, but the way of the wicked will perish.

It is easy to recognize the orientation of time to action. After the warning of an unformed time of the godless, that is, sinners and mockers who lived only for the day, a clear instruction for action followed. Through re-

---

13. This occurrence is the only place in the text where 'et stands for the concept 'olam. We will show below whether the translation of "eternity" is suitable.

flection, which meant a special form of meditative reading, the devout shall have internalized the law of the Lord. This presupposed that there will be a corresponding realization in action in everyday life. Success in this active formation of time was not to be long in coming (vv. 3b-d: "which yield their fruit in its season"). The devout could enjoy the fruits of their efforts in a social context. A collective orientation built a common social world (v. 5b), the community of the just, which was defined over against those (v. 4, the godless) who did not satisfy this maxim. Their lack of action could not build a formed time, and since there was no common social time, they were isolated like chaff, which the wind would blow away. It is interesting that the psalm postulated a dependence on God for human actions. Without religious usage and practice there was no reference to God. God's judgment revealed the inactivity of the godless, and sinners do not stand in the judgment. However, the devout could reckon with God — and the divine structuring of time? — for on their way, they were connected with God. The knowledge of God meant more than merely perceiving or knowing; it included sharing and walking together with God.

The second concept for time, *'olam*, is usually translated as eternity. However, it had a more particular meaning. It could mean the most distant time, such as, primarily in later texts, the time of the world, in the sense of a cycle closed in on itself. In the preexilic texts of Micah, Amos, and Hosea it was hardly related, for it suggested that in this early time the horizon of time was only weakly distinctive. Only in the prophetic texts of Second Isaiah, in connection with the eschatological action of Yahweh and its universality, did the claim of this concept gain a new dimension. We will come back to this point later.

The third concept for time was *roega*. It marked a short period, a moment.

The lexical results of the words for time do not allow us to draw any conclusion on complexes of ideas that were somehow disposed to recognize time as such. *'Et* marked points of time that were especially qualified in content. *'Olam* was purely a concept of relationship. *Roega* was a technical term for very short intervals of time. Their usage can be expressed differently: the Hebrew language did not approach the phenomenon of time through *abstract thinking*. Rather, all investigations agree that the specifically Hebrew understanding of time was always articulated in the context of certain events or human activities. Corresponding with our thesis, the Hebrew understanding of time was expressly oriented to action. This statement agrees also with the lexical results of the Hebrew word for action (*asah*), which is the most frequently used word in the Old Testament. In addition, the Israelite religion brought religious salvation into connection with human action, for example, in Psalm 1.

b. Endogenous Time of Nomads We can only speculate about the experience of endogenous time. We may assume an analogy with the mythic experience of the early high cultures. As already described, the latter were exposed again and again to situations of existential stress, which served as an external stimulant for development. Over the course of time, with an asynchronism between endogenous and exogenous time and the scarcity of resources, it is important to establish that we have met two important factors. These factors could have contributed to a more immanent basis of an early forming of a consciousness of time, and they could have contributed to historical thinking. However, there is a question as to whether this immanent basis is sufficient to explain the historical consciousness that actually existed, as well as the specific experience of time by the Israelites. At this point we have touched the third aspect of our tri-polar structure of time.

c. Transcendent Time, the God of the Fathers The religious experience of the patriarchs of Israel took place outside of the Israelite faith in Yahweh. Ancient deities, *numina*, whose names have been handed down fragmentarily to us in the Bible, are characteristic of their religious world. All occurrences of them are named here:

'Abiyr Ya'akob = the Mighty One of Jacob, cf. Genesis 49:24 [NRSV]
Pachad Yitshaq = the Fear of Isaac, cf. Genesis 31:42, 53 [NRSV]
El Shaddai = God Almighty, cf. Genesis 17:1; 28:3; 35:11; 48:3; Exodus 6:3 [NRSV]
El 'Olam = the Everlasting God, cf. Genesis 21:33 [NRSV]
El Beth-El = the God of Bethel, cf. Genesis 31:13; Genesis 35:7 [NRSV footnote q]; cf. also Genesis 28:18-22

Reminiscences were handed down of this pre-Israelite form of religion or forms of religion, and they are also located in Genesis 31:19; 46:6; Joshua 24:2: "Thus says the LORD, the God of Israel: 'Long ago your ancestors — Terah and his sons Abraham and Nahor — lived beyond the Euphrates and served other gods'" [NRSV]. What was meant by these deities? What do they have to do with our question? The answers become more difficult for us to find because the lives of these deities have not been handed down clearly by the tradition. Rather, the texts that report about these other deities have been reworked many times, and certainly the ancient names of more gods were deleted by scribes. In addition, these stories of the patriarchs followed a theological tendency, which in the first century after the origin of the sagas was interwoven with certain other theological tendencies that had a current theological influence connected with them. This process especially became clear with the name El Shaddai. The texts in which this name occurred were all indebted to the so-called Priestly tradition, which was current about 600 B.C., long after the patriarchs. In these texts, the original meaning was covered up by the theological influence that followed. Nevertheless, something of the peculiarity of the names remained that opens the door for supposing they refer to deities. Two of the deities were personal, and they were connected with a certain quality of experience. We can suppose they got their names when they first appeared. The name of the remaining three deities began with El, a common Semitic name that was used in other ancient oriental religions. Even the Old Testament has preserved such a deity's name in the name El Elyon, the God of the heathen King Melchizedek (cf. Gen. 14:18ff.). The significance of El Shaddai hardly amounted to more.

El Beth-El dealt with a cultic place for God. Many such cultic places were required in the time of the patriarchs. *Numina* were honored at them and gave them local significance — these deities were therefore static and bound to space. Maybe they were connected with certain people or clans, and maybe they were also dynamic and bound to time. Albrecht Alt (1970) presented this interpretation of the *numina* after an extremely careful study that compared extensive religious-historical materials. The study of religion speaks here of the spatial epiphanies of deities (Moltmann 1964, 85ff.), in contrast to the nomadic deities who were related to time.

We observe in the early Israel of the patriarchs the extremely interesting displacement of spatial epiphanies of the deities by the nomadic deities who are related to time. The process of displacement then was mirrored in the theological reworking of stories of the origin of cultic places, so-called cultic etiologies. The stories went back to the spatial epiphany of a god, and then they were taken up secondarily by the God of the nomads, who had a monopoly on time and history. Where the spatial cult of deity was involved, we are in the fortunate situation of being able to fall back on a text in which this God was intended, even though the name did not occur *expressis verbis.* The subsequent theological reworking spared the remaining text, and therefore the reader is offered an unadulterated story in its entire archaic splendor.

In Genesis 28:21 we suspect the handwriting of later theological redactors, who replaced the ancient name of God with the trusted and accepted name of Yahweh. In this well-known story of Jacob's seeing a ladder in a dream, we read in verses 16-22 what Jacob did after waking up. Jacob concluded a covenant with God by promising that he would serve God if God would provide for his well-being. That is, there was a *relationship* between God and the *time* of a human being. Genesis 31:13 and 35:7 note that God kept this covenant promise. We thus already find on this archaic level of faith in God a special relationship of God to the time of humans — more exactly, to their future, to their *open* future. This aspect would be seen ever more clearly in Israel's history.

This relationship of God to the future of humans can be seen especially as it was expressed in the form of a transitional figure between nomadic times and this incipient people. In Abraham this new horizon of the future and the oneness of time took concrete form. The special relationship to time by the new God Yahweh was expressed in the form of an invitation to Abraham to break with his past. It was affirmed in the form of the promise of three blessings for his future. The text of Abraham's call, which has been reshaped by the redactor J, in Genesis 12:1-3 reads:

1 Now the LORD said to Abram, "Go from your country and your kindred and your father's house to the land that I will show you.
2 I will make of you a great nation, and I will bless you, and make your name great, so that you will be a blessing.
3 I will bless those who bless you, and the one who curses you I will curse; and in you all the families of the earth shall be blessed." [NRSV]

The connection of blessing with promise in this text is interesting. This connection of the two religious areas, which had been separated to this point, was absolutely new and original. Furthermore, the promise was definitely recent, and it already presupposed the new concept of God. It is interesting in this respect because the idea of a magic-mythic blessing as an immediate powerful effect was overcome in favor of a promised personal way of life (Westermann 1981, 173). It was the final step that crossed over the bewitchment of spatial, mythic immediacy, and it opened time as the dimension and perspective of faith (Abraham was the father of faith [Gen. 15:6; Rom. 4; Heb. 11:8-10]).

Summary   If we now look at our tri-polar key *uno aspectu,* then we can establish the following as a result: in contrast to the other two ancient oriental cultures, in Hebrew society the tri-polar system was already so severely disturbed at the mythic stage that it could not become stable. Rather, this system constantly drove out the mythic stage. The stimulus for this development was the asynchronism between transcendental time and endogenous and exogenous time. A characteristic of the Hebrew perception of time emerged as the *openness of time as a dimension and perspective of faith.* Nevertheless, the nomadic Hebrews did not manage to attain the linear structure of time.

We have discussed in detail the effects of various combinations of our tri-polar system on the mythic-cyclic level. In the end, all three cultures remained in the bewitchment of the mythic experience of time, although in all of them impulses for overcoming it were recognizable. We turn now to a culture that succeeded in the transition to the rational-linear and the mystic-holistic structures of time: ancient Greece.

Overcoming of Mythic Experience of Time by the Greeks

a. The Mythic-Cyclic Experience of Time by the Greeks in Early Ancient Time
The experience of time by the Greeks was strongly influenced by the danger of human life. As a seafaring people, the Greeks knew about the deadly, unpredictable dangers of the sea. Dreadful epidemics, such as the plague, were portrayed in detail in the *Iliad.* Also, the Achaeans were conscious of many threatening powers: drought, catastrophes, bad harvests, and earthquakes. The ancient Greek experience of time was thus not influenced by stable earthly cycles, as in Egypt, but by the experience of transience and the deadly power that "produces everything and takes everything away again, long and uncountable time" (Sophocles, *Aias,* 646-47).

Exogenous natural time appeared in Greek culture in a twofold form: At first it appeared as *chronos.* Time as *chronos* was "long and uncountable time," which produced the changes of beginnings and disappearances, of luck and misfortune, of well-being and woe. These changes were cyclic, but not in the strict sense of the exact repetition of phases. They were cyclic in the sense of the unpredictable tendency of ups and downs in the existence of life and death. Life and death formed two irreconcilable spheres: the sphere of god, which was immortal, and the sphere of the mortal. Homer compared the human race with a flash in a pan, and with leaves that fall from trees in autumn. Yet the great poet Pindar saw human beings as being "nothing" (*Pythien* 8.81): "The nature of a day: what is it, and what is it not? A human being is a shadow in a dream. But if god-given glory comes, a bare appearance lies on men and mild existence."

Natural time meant a change of luck and misfortune, an eternally rolling wheel, which drove its unpredictable game with mortals. The other concept of time was called *aion* = lifetime, the natural span of life. Each creature had its naturally measured *aion*. The cosmos had its *aion*. The Greek concept of eternity later developed from it. Here, time was seen as a specific measurement that made possible the structures of life. As distinct from the New Testament concept of time, these *aion*s were not comparable with one another. Time as *aion* avoided the measuring rationality and capriciousness of human beings.

Nevertheless, the Greeks were able, presumably through influence from the Babylonians, to understand the temporal order of nature. However, their realization of it in socially binding times in the form of a calendar had to do battle with various difficulties in Greece's particular political organization.

Since the first millennium B.C., Greece's polycentric structure led each city-state to develop a calendar on the basis of the lunar cycles (Seleschnikow 1981). In order to gain agreement with the seasons, intermittently they introduced an added month. The beginning of the year was the summer solstice (*hekatombaion* = July). Furthermore, there were different cycles for leap years. The best-known Greek calendar was that of Meton with a nineteen-year cycle (19 sun years = 235 moon months). It is remarkable that probably this cycle in the calendar of the Meton had been converted into a mechanical simulation. A copy of such a "human clock" was found in 1900 on the wreck of a Greek ship near the island of Antikythera, and in 1974 it was examined scientifically (Whitrow 1991).

Already the border of rational-linear time was exceeded, and the Western measurement of linear time by clocks began in the late Middle Ages. From Meton it was only a small step until time was counted. How important to the Greeks the measurement of time was may be seen on the occasion of the 82nd Olympiad in 432 B.C. In the center of Athens a tablet *(parapegma)* was established as a reminder of annual astronomic appearances.

At first there was no counting of years in ancient Greece. The years were named for leading public servants. Since the fourth century B.C., the years were numbered after the Olympiads, and that was the practice for seven centuries. It continued well after the abolition of the Olympiads in 394 A.D. by Theodosius I. The Greeks have always countered the unpredictable power of *chronos* with social timers. Already with Homer the times of the day were marked by human works that were begun or finished, such as harnessing (= mornings) and unharnessing (= evening) the oxen. Duration was expressed by putting it in relation to a generally known performance, for example, covering a known distance.

The strongest social timers were cultic celebrations, and especially the greatly respected festivals in Olympia in the land of Eli in the huge temple district of Zeus. Since 776 B.C., they were held regularly every four years, following the Greek tradition, and from then on the Greek calendar was dated by "the Olympiads." At the time of the games, which lasted five days, all wars and violent disagreements ceased throughout Greece. The Greeks, therefore, succeeded in gaining an *arrangement of exogenous time* that tried to resist the powers of war and transience.

The *experience of endogenous time* by the Greeks was expressed in myth, which was an especially rich method of expression for them. As we recognized in the Greek myths, the ancient endogenous time was synchronized to the exogenous time of experience: the external experience of time as *endangerment* corresponded to the inner experience of time as *being at the mercy* of the unavailable powers of the deity. Over the sanctuary of Apollos and Dionysus in Delphi stood the sentence: Know thyself! That was not meant as an invitation to psychological self-exploration, but rather as a reminder: "Be thinking always and everywhere that you are only a human being and nothing more!"

That saying became particularly clear in Greek tragedy: people were not allowed to transcend their earthly sphere. As soon as he rose into the sphere of the divine, the divine world order inexorably beat the "hero" back and threw him into the abyss of undoing. The inner perception of time was pessimistic and cyclic.

Corresponding to the mythic-cyclic stage of consciousness, the religion of the ancient Greeks was to a large extent polytheistic. The goddesses and gods of the Greeks were not transcendent but rather components of nature, *physis*. The variety of nature was reflected in the variety of deities. Each deity represented a certain aspect, as it were, or a certain dimension of existence. They were not omnipotent either, but they were very much more powerful than humans. However, they were put beneath the final powers of "necessity," of "providence," and of "fate." The gods had their own "time," their own *aion,* and it was precisely here that a bipolarity appeared that is so important for the understanding of Greek culture. It is the bipolarity of the clear, pliable, limited "time" of the Olympian gods, and it was present especially clearly in the form of the god Apollos and the exhilarated, formless, unlimited "time" of the god Dionysus.

b. The Experience of Rational-Linear Time by Xenophon With Xenophon of Kolophon (born 565 B.C.), the "storm petrel of Greek Enlightenment" (W. Capelle), for the first time a rapid breakthrough appeared in the mythic-cyclic experience of time in the direction of linear time, and it was historical in the real sense. This sentence was handed down from him (B 18): "God did not reveal everything to mortals from the beginning, but only by searching do they find the better, little by little."

A completely new perception of time was expressed in these words: History was felt to be an ascending line, which led to ever greater realization. Later, Aristotle conceptualized this rational-linear understanding of time in greater detail. It was the educational pathos of *progress,* which was reported here for the first time in Europe. What were the causes?

The story of Xenophon is very moving. At twenty-five years of age he left his hometown of Kolophon on the coast of Asia by escaping from the Persians, and he sought his fortune far away. Various wanderings led him to Sicily and Malta, and maybe also to Egypt. He found his calling as a rhapsodist, that is, as a reciter of Homeric poems. He thrashed around through life, and finally he found a permanent home in Elea, on the western coast of southern Italy. Obviously, the fullness of these new experiences and impressions of different cultures and their different structures of time moved him to think comparatively and critically. Time-honored patterns of thought and experience were questionable to him, and that awakened independent thinking in him. In an intercultural comparison, mythic-cyclic thinking appeared contradictory to him, and so he emancipated himself from it. His own experience of time changed. Here we come to endogenous time.

Xenophon dissociated himself from the keenest of the traditional standards for structuring social time, which standards were established by the cults and the Olympic games. Over against the idolatry of the Olympians and their athletic abilities, he placed his own wisdom. Only through it would the well-being of the city be promoted, he said (B 2):

> If one carries off the prize in the customary competition and would be highly honored by his fellow citizens, still he would not be of as much value as I am. That's because my wisdom is better than the raw strength of men and horses. However, each cult lacks every inner legitimacy. Therefore it is completely unfair to value brute strength higher than precious wisdom. And if it appears that one is proficient among his compatriots as a boxer, or in the pentathlon, or as a wrestler, or by the speed of his feet — whatever in the competition of these men is worthy as a sign of particular strength — still the well-being of the city would in no way be promoted by it.

The self of the "I" experienced the time of wisdom and progress. For Xenophon, this realization of the self turned into a critical potential, which itself was asynchronous with the mythic-cyclic patterns of time. A massive criticism of religion arose from this critical, endogenous potential. It was a criticism of the traditionally experienced time of the gods.

Xenophon expressed damning criticism of the anthropomorphic polytheism of his people, that is, the world of the gods of Homer or Hesiod. His alternative was — just as with Echnaton's reform, which was caused by a strongly self-conscious "I" — a strict, ethereal monotheism (B 23): "There is only a single God who prevails as the greatest among gods and people, similar neither in appearance to mortals nor to thoughts." Of him is now said (B 24): "Completely he sees, completely he thinks, completely he hears." And further (B 26): "Always, he pauses at the same place, which is not moving at all; it is not proper for him to go here and there." God was, as it were, the guarantor of the unity of time, an absolute time, within which could occur something like cultural *progress,* a progress that reformed critically the old mythic-cyclic tri-polar structure of time entirely.

c. Parmenides' Mystic-Holistic Experience of Time   Parmenides was a student of Xenophon, and he also lived in Elea (lower Italy). He is rightly regarded as the founder of Western metaphysics. In his great teaching poem, an experience of eternity was revealed. This experience constituted a negation and a disavowal of flowing, passing time. Xenophon's experience of rational-linear time was transcended by Parmenides in an experience of "timeless," "mystic" time. The "now" was experienced as the point where all truth came together, and it was where everyone experienced true time and all true being.

To be sure, Parmenides was politically active in his hometown of Elea — he is supposed to have invented and established important laws — but his relationship to external time was impressed by strong *flight from* and *contempt for* time. Because of a Pythagorean influence, he led a secluded, virtually "monastic" life. Exogenous time increasingly turned into an abhorrent object for him, and thus he talked in fragment B 12 of love, mating, and birth as something "horrible." According to him, everything temporal was negative and vanity.

Parmenides was completely inspired by the mystic-holistic experience of an absolute, inexpressible *unity* of being and time. Hermann Fränkel wrote: "Thus there is a weighty argument in favor of our assumption, that Parmenides personally experienced the *unio mystica* with the true being" (Fränkel 1931, 418). In his teaching poem, he described in poetic language how he was transported into an unearthly region of light, and there he saw the truth: that being is, and nonbeing is not. All multiplicity, all corporality, and all temporality belonged in the sphere of appearances. In these appearances, however, true being appeared to be broken. That is, the true being was eternal and immortal, and it was the eternal present. The eternal present was the one and all-inclusive form of endogenous time for Parmenides.

What was learned from Parmenides was that the unity of time was transcendent, divine. It was that which was undoubtedly always implied in all the talk and experience of "time," "times," and "modes of time": the cohesion and the unity of past, present, and future, which alone justified calling these three modes "time." This unity of time guaranteed the cohesion of the world.

Parmenides now made the great attempt to conceive this unity of time in a certain mode of time, namely, the present. There was a good reason for this attempt: the present has priority over the past and the future. This priority consists in the fact that the present alone "is." The nature of the present is manifestation. Also, past and future "only are," providing they participate in the nature of the present through the interlacing of the modes of time. The present was the "threshold" (B 1.2.11ff, cf. Hesiod, *Theogony* 744-45) that all who wanted to enter into the light of truth and being had to cross. It was obvious, consequently, to see the unity of time in this unchangeable "threshold of being." Only so could the unity of time be understood as reality, that is, understood as presence. With Parmenides, for the first time there appeared a concept of "eternity," an idea transcending that of a very long or infinite duration. "Eternity" was the negation of past, present, and future as we experience them directly. The difference in the modes of time blended into an understanding of the unity of time: "Thus origins fade away and offenses are forgotten" (B 8, 21). Eternity dropped out of time. Therefore, up to the present time the metaphysics of Parmenides has been understood as saying that truth is timeless. With Parmenides, this view has the radical consequence that time was simply banished from the area of thinking (B 2). The tri-polar structure of time collapsed with Parmenides into an experience of mystic-holistic unity.

d. The Integration of the Greek Experience of Time in Plato The radicalism of Parmenides could not persevere in Greek philosophy because it remained responsible for the *explanation* of time, that is, from whence it came and how it was obtained. Plato's philosophy can be understood as an extensive synthesis of the two contrary beginnings of Parmenides and Heraclitus. He developed a theory in which both eternity and time are combined in a relationship.

This theory was the first basic design for European philosophy that comprehended time and eternity. We sketch it briefly here. It appeared in *Timaeus* 37c–39e. Plato portrayed there the perfection of the universe (*Timaeus* 37c6–d8) in the context of its myth of the creation of the world:

> When the father and creator saw the creature which he had made moving and living, the created image of the eternal gods, he rejoiced, and in his joy determined to make the copy still more like the original; and as this was eternal, he sought to make the universe eternal, so far as it might be. Now the nature of the ideal being was everlasting, but to bestow this attribute in its fullness upon a creature was impossible. Wherefore he resolved to have a moving image of eternity, and when he set in order the heaven, he made this image eternal but moving accord-

ing to number, while eternity itself rests in unity; and this image we call time.[14]

The *a priori* of Parmenides had been overcome: eternity was no longer merely the negation of time, but rather it was the original image of time. For this purpose, Plato took up the concept of *aion*, which was originally in Greek "lifetime," "entirety of a life span." He filled it with new meaning, and he understood it as the eternal *aion* of the cosmos. The *aion* was not flowing like time, but it "persists in the one." Time was its image *(eikon)*. This ambivalent nature of time can be stated this way: It was *homogeneous (continuum)* because it always appeared to be the same, namely, as eternity. It was *flowing* because the unity in it only appears, and therefore it was not actually present as an existing present. Consequently, according to Plato, time was purely the appearance of pure being.

Now time was the existing unity in a flowing succession. However, the same was true also of the sequence of numbers: there was a unity in this succession because each number was a countable part of the growing sequence, and therefore (as a form) it was an uncountable unity in itself. Time and number therefore were isomorphous, and that was why Plato could say time "progresses in numbers." This view explained their mathematical ascertainment by means of numbers and measurements. Consequently it asserted the mathematical ascertainment of nature — a deep thought that justified the possibility of the philosophy of nature in *Timaeus*. This theory of time overcame the *a priori* of Parmenides, which remained bound to its base, as the following section showed (*Timaeus* 37e4ff.):

the past and future are created species of time, which we unconsciously but wrongly transfer to the eternal essence; for we say that he "was," he "is," he "will be," but the truth is that "is" alone is properly attributed to him, and that "was" and "will be" are only to be spoken of becoming in time, for they are motions, but that which is immovably the same cannot become older or younger by time, nor ever did or has become, or hereafter will be, older or younger, nor is subject at all to any of those states which affect moving and sensible things and of which generation

---

14. *The Dialogues of Plato*, trans. Benjamin Jowett (Chicago: Encyclopedia Britannica, Inc., 1977), p. 450.

is the cause. These are the forms of time, which imitates eternity and re-volves according to a law of number.[15]

Two things are remarkable here: First, the devaluation of the temporal modes of past and future by Parmenides was fully accepted. Second, "time" was said to be "revolving," and this view was the necessary consequence of the Platonic beginnings. If the eternal present were to appear in the time of the world, it was a world that the Greeks were convinced was following the circularity of time. Only in a circle could time be uniform. Eternity always appeared equally near in this *aion* of a finite world. We have here a complete model of the metaphysical understanding of time and eternity, and it has worked decisively and persistently in Western philosophy and theology. The three stages of consciousness of the mythic-cyclic, the rational-linear, and the mystic-holistic experiences of time were summarized and integrated in a higher unity.

Summary   Greek culture produced all three stages of consciousness of time. It is interesting that the transition from the mythic to the rational-linear experience of time was produced by an experience of extreme asynchronism between endogenous and exogenous time, and it was connected with the rational criticism of the mythic belief in gods. We saw the same thing in the Mesopotamians and Hebrews. In contrast to these two ancient oriental cultures, ancient Greece had taken a leap forward to the *linear structure of time*. Also, the following stage can be interpreted as leaving the rational-linear structure of time for the mystic-holistic one, as a by-product of an asynchronism between endogenous and exogenous time. Parmenides' distance, indeed a retreat from the world and the linear formation of time, brought to him the linear rhythm, so to speak, and it offered the opportunity to open the linear form of time to the mystic-holistic one.

Concluding with Greek society, we have now met many possibilities of combinations of the *entire* tri-polar system. We now want to turn separately to the rational-linear and mystic-holistic experiences of time. We particularly want to investigate the development of the concept of rational-linear time since in our Western society this concept has been the dominant paradigm up to the present.

15. *The Dialogues of Plato*, p. 450.

The Experience of Rational-Linear Time in the West

The experience of rational-linear time in this scientific and technological world, which sees time as a linear parameter that can be read from the hands of a clock, is an exogenous experience of time. This view dominates all processes in technical civilization. All the technical appliances that determine our lives are built so that the processes in them can be portrayed on a straight time line. Humans do not experience this rational-linear structure of time as a predetermined development discovered in nature, which is usually cyclic. Rather, it has been constructed by the culture formed by human beings. It is primarily a human construction, a performance of human culture, which only secondarily overlays the cycles of exogenous and endogenous time cycles. By the spiritualization of humans, at least in Western society, this view becomes second nature in experience. We think of and experience time as straight, and it runs from the past through the present into the future. From whence comes this linearity of time?

In the ancient cultures we examined, we saw that the embryonic linearity could be interpreted primarily as a reaction to experiences of asynchronism between endogenous and exogenous time. Nevertheless, the experience of linear time was not successfully durably established in social time in ancient cultures. However, not only has this experience of time been brought to the West by its selective flashing through the mind, but it has also helped without parallel in a triumphal march up to the crisis of the present. What are the reasons for its origins, and for its power to endure in organizing life for centuries?

As we saw, the experience of linear time is always joined with a higher level of subjectivity and stability of the "I." In the search for reasons for the origins of the experience of linear time, we will not go wrong if we look into the intellectual history of Europe for a thrust toward individualization, and for the conditions that constitute it. In addition, we are already familiar with the experience of the desynchronizing of endogenous and exogenous time, and we will direct our attention to factors that favor the process of subjectivization. These must be factors in which humans are thrown back on themselves. Such experiences deny humans an orientation of life by *external* stabilizing factors. Only then do the necessary motivation and the necessary distress put pressure on people to create and to build an *inner* counter-world. The desynchronism we described only included a special case of the general experience of insecurity. If we keep a lookout for the coincidence of the experience of external insecurity and the thrust of inner subjectivization, we will quickly strike paydirt. In the fourteenth century in Europe both factors coincided in an extreme way.[16] The depth and penetration into virtually all areas of life over three generations made it reasonable that the experience of subjectivization could produce such a far-reaching effect. Many areas of life found an empowering of their temporal performances. Here are some examples:

The social order was shocked at this time by the decline of knighthood, which had prevailed up to that time, and by the ascent of the middle class in the rapidly growing cities, particularly in upper Italy. The Catholic Church entered into uncertainty by its move of the pope from Avignon and the escalating Inquisition. We can hardly overestimate the devastation of the plague in Europe, when from approximately 1348 to 1352 about 30-50 percent of Europe's population fell victim to it. Depression in the economies accompanied the beginning of the transition from an agrarian subsistence to urban capitalism. The collapse of the boldly constructed Gothic cathedral of Beauvais in 1284, the minimal continuation of construction in 1324, and finally the failure to complete this special construction could be interpreted symbolically as a passing of the zenith of the Gothic style in the fourteenth century. It happened in the paintings of Giotto (1266-1337), when Byzantine painting, which created an impersonal and idealized picture of a saint, was replaced by an emphasis on individual features and their respective arrangement. In music the form of the French Ars Nova began in about 1320, as well as the Italian madrigal style of music that was more oriented to time than to the sounds of space. Worldly, middle-class music became generally accepted with its expression of situations of the middle class.

Also there were changes in the world of the mind. The so-called controversy over universals, the question of the ontological status of generalities, or general concepts, resolved into the unquestioned participation of the individual in the generalities (Plato), into the independence of the individual (Aristotle), and into unique, unrepeatable particularity (nominalism). Consider this: already in 1317 — about three hundred years before Descartes! — in his commentary on *The Sentences*, the scholastic

---

16. The significance of the fourteenth century for the intellectual history of Europe has often been mentioned (cf. Huizinga 1957; Blumenberg 1974, 1981).

Peter Aureoli questioned the *certainty* of empirical, theoretical, and moral perception. This tendency of making things uncertain by questioning was picked up by all theologians who were inclined towards nominalism, and they carried it further until the fifteenth century. How strongly this nominalistic uncertainty worked on people can be measured by the fact that the study of nominalism was prohibited in the fourteenth century at many universities: 1323 in Oxford, 1340 in Paris.

Finally, the great rational syntheses of scholasticism broke apart. Religious certainties became uncertain. With Thomas Aquinas, natural and supernatural, faith and reason, divine mercy and human action, Scripture and tradition were still synthetically related to one another. William of Ockham dissolved this synthesis at the latest in the fourteenth century. Faith and reason separated from one another, and the principle of twofold truth arises. Knowledge should henceforth be acquired and defended by experience — the first scientific centers originated in the fourteenth century in Oxford and Paris — and not by revelation, as if it were already available. Finally, this disruption did not stop at the picture of God. Thomas Aquinas's picture of God was oriented toward *existence*. William of Ockham's picture of God assumed *work*. The gap between these views of God resulted in different views of life. The first view consisted of a predetermined order of being and salvation, and life was given up to contemplation. The second view consisted of the active formation of faith, which recognized the questionable order of perception, being, and salvation and replaced them with its subjective perception and forms of action. Not without reason was Ockham a voluntarist.

How strongly the dissolution of externally arranged predetermination led to subjectivization, and thereby to religious spiritualization, can be seen in the interpretation of the sacrament of penance in nominalistic circles. The grace of the sacrament of penance could only become effective, according to Ockham, if it were supported by a strong inner impulse of remorse *(contritio cordis)*. The traditional conception that was oriented to an external sacramentalism of "beginning remorse" *(attritio)* was no longer enough for him. In addition, the nominalists radicalized the insecurity of the experience of the state of grace. According to their voluntaristic picture of God, and in accordance with his absolute power, God could rescue sinners and damn the just. God was no longer a refuge for human security, but rather the Initiator of constant threats to all human certainties. This religious experience of insecurity turned into a movement to leave the closed world of the factual and to turn to the open horizon of possibility. The dimension of possibility was faith. However, the area of faith understood this way was time.

The uncertainty, the subjectivization, and the tendency toward isolation were now reflected also in the philosophical discussion about the concept of time. Thomas Aquinas (1226-74) suppressed Plato's ideas in the thirteenth century, and Aquinas's acceptance of Aristotle had prepared the way for the acceptance of Aristotle's definition of time as the *measurement of movement*. The philosophical problems connected with this definition of time were vehemently discussed in the generation following Aquinas. Those to be named in particular are: Henry of Ghent (1217-93); Dietrich of Freiberg (1240-1318/20); Peter Johannis Olivi (ca. 1248-96); Heinrich of Harcley (1270-1317); Wilhelm of Alnwick (1270-1333); Walter Burley (1275-1344); Peter Aureoli (ca. 1280-1322); Nicolaus Bonetus (1280-1343); Johannes Buridan (ca. 1300-1358); Albert of Saxony (1316-90), founder and first rector of the University of Vienna; Nicole Oresme (1325-82); and Marsilius of Inghen (1340-96), founder and first rector of the University of Heidelberg. At the end of this discussion stood the total subjectivization of time and its identification with movement by the nominalist William of Ockham (ca. 1295-1349, died from the plague).

This new relationship to time was expressed in a new language: R. Glasser (1936) has proven that in fourteenth-century France, the leading country intellectually at that time, many new verbs for temporal action came into use. These words demonstrate that the conditions for the process of subjectivization — the breakdown of meaning and orientations that mediate order — were present in all areas of life across the board, so that the concept of linear time could arise from a strengthened subjectivity. This massive destabilization was probably not enough to provide the thrust to establish this individualization permanently, for there are always crises in human history, crises that have objective power. The perception of linear time was not independent of the subject, so to speak, and constantly required internalization.

The invention of the mechanical clock at the end of the thirteenth century, and its introduction to the public in the fourteenth century, guaranteed the regularity of time and its social implementation, as well as the permanence and the triumphal march of linear time. We now want to trace this process of establishing linear time in individual lives — coupled with the thrust toward subjectivization and individualization. We want to begin with the Aristotelian prelude to linear time in Greece, as a prerequisite for the conceptual establishment of linear time in the late Middle Ages. Philosophically, our present-day understanding of linear time may be seen as a synthesis of the Aristotelian idea of time and its theological digestion in the Middle Ages.

As Georg Picht presented impressively (1980, 362ff.), in Greek philosophy the unity of time was understood as the identity of the eternal presence of being. Understood as unchangeability, the modes of time — past, present and future — could not be derived from the essence of time. Since Aristotle, they fade out, and time has been conceived as a circular line. This view followed from three axioms, which the Greeks considered to be valid:

1. There was an unchangeable identity, the absolute being.
2. The world was final. It was a sphere.
3. The unchangeable identity appeared in the world through the medium of time.

If the unchangeable identity appeared in a finite world, which was spherical, then time had to be thought of cyclically. Up to this conclusion Aristotle followed his teacher Plato (see above). However, he consciously distinguished himself from Plato in that he dissolved the connection between *aion* (world time, one's own time, eternity) and *chronos* (flowing time). He did it in order to dedicate himself to a detailed investigation of *chronos*. In Book IV of *Physics,* chapters 10-14, we find his basic treatise on time. Aristotle asked about the nature of time in the framework of the teachings about movement. There he defined time as "a measurement of movement in the view of the earlier and later" (219b1f). His access to the understanding of time was no longer cosmological and astronomical, but purely physical. He brought time, as it were, down from heaven to earth, and he viewed it as "something in motion," as "a measurement of motion" (220b32f). Time was portrayed by the middle concept of mobility in space, or, could we even say, "spatialized." In this way, time became only a metaphor for a circle, but it was thought of geometrically as a quasi-circular line. Here was the basis for the concept of linear time in modern physics!

The experience of rational-linear time in the Christian West and its civilization presupposed the essence of time that Aristotle had developed in Book IV of his *Physics*. Here, the figure of a linear parameter was created, which annihilated the difference between past, present, and future.

Aristotle said time in the difference of its modes "either has no existence at all, or only barely, or has become blurred" (217b32f). Consequently, we could say that the foundation for the typically Western concept of time was already laid in classical antiquity. Its characteristic Western impression received its shaping through its relationship with specifically Christian ideas in the late Middle Ages, after Christian theologians, mainly Thomas Aquinas, had absorbed Aristotle in the high Middle Ages.

The Thrust toward Individualization in the Fourteenth Century as the Basis of the Perception of Rational-Linear Time

a. Time and Theology The god of the Greek philosophers was one who was absolute, the absolute reality, and who embodied the concept of the *apeiron*, to which infinite could not be exactly applied because the divine was closed on itself and was complete in itself. Plato's "idea of the good" was the quintessence of the good, in that it was the form of forms. The concept of "unlimited" or "infinite" had nothing in common with this sphere of the divine.

In the Jewish-Christian area, a quite different conception of God appeared: the God of Israel, who was revealed to Moses at the burning bush, was not "the absolute being." God resisted each possibility and attempt to be expressed in thinking. God reveals God, rejecting and inclining toward at the same time, with the words: "I WILL BE WHAT I WILL BE" (Exod. 3:14 NRSV, footnote e) The LORD again and again proved to be the Lord of history in inexhaustible saving power for the people. The God of the Old and New Testaments is omnipotent. "For God all things are possible," says Jesus (Mark 10:27 NRSV). In the Gospel, there was a priority of this possibility for the real (Jüngel 1972, 206-34). The God of the Bible is the "coming God," who makes the irreversible process of history possible from the future. "'I am the Alpha and the Omega,' says the Lord God, who is and who was and who is to come, the Almighty" (Rev. 1:8 NRSV).

The God of the Bible is not absolute reality; rather, God is the inexhaustible possibility and feasibility of all being. God does not want this creation to be lost but rescues and always redeems it. In the Gospel there is less talk of the eternal reason or wisdom of God than there is of God's will (cf. the third petition of the Lord's Prayer: "Thy will be done!"). This view was newly discovered in the radical discipleship of Francis of Assisi (1182-1226). In his Order, the Franciscan Order, with the rejection of the static

Platonic picture of God, nominalism developed that taught, among other things, about "God's infinite power" *(infinita potestas)* and therefore presented a dynamic picture of God. This *potentia dei absoluta,* however, was an expression of the category of possibility, which implied a linear structure of time that was open to the future. From the objective side, from the side of God, so to speak, the prerequisite had been created for the rational-linear conception of time. God was not only the Lord of time, which Augustine already knew, but also the source of the possible new time.

This view had an effect on the concepts of the world and time. We find it most clearly formulated by Nikolaus of Kues (1401-64). In accordance with the old philosophical statement that the divine image appeared in the world (see above), he concluded:

- God has unlimited power and is infinite.
- God has created the world as a counterpart and an image, and time was created as the counterpart and image of God's eternity.

It followed, then that the world was not a *finite* sphere — as with Aristotle and his Christian followers — but an *infinite* sphere, which did not have a definite center and had no periphery (Jaspers 1964, 102-21). As Aristotle had already explained, for Cusanus time was an attribute of this moving world. Aristotle's line of a finite circle turned into a line of an infinite circle, and therefore a straight line! Connected with the old biblical idea of holy history, which had an irreversible direction, the modern concept of time arose: rational and linear. "So, the change in the concept of God leads to a transformation of the form in which the unity of time is introduced. The thinking of modern times presents time in the picture of an infinite straight line. The unity of time is uniform and one-dimensional" (Picht 1980, 364).

b. Time and Philosophy This new, divine, and objective structure of time had its counterpart in the structure of subjective human experience. Seen philosophically, it happened as philosophy slowly emancipated itself from theology in the high Middle Ages. With Thomas of Aquinas it had the function of serving theology, and it was true that philosophy was a handmaid of theology *(ancilla theologiae)*. For the first time philosophy gained an autonomous role with William of Ockham in his strict separation of reason and faith. Henceforth, philosophizing was only a measuring rod and law. However, this led to further individualization of philosophical thinking. This tendency was shown in the handling of different philosophical problems, as in the controversy over universals (indicated above), and also in the handling of the problem of time.

The controversy over universals was about the theoretical realization and ontological role of general concepts. Were universals as such in themselves, such as beauty is in itself, as Plato thought *(ante re)*, or were they only in individual things *(in re)*, as Aristotle claimed? William of Ockham developed a revolutionary view to overcome the handicaps of the ancient philosophical tradition. Quite self-confidently *(ego dico)*, he thought that all the general concepts had no ontological meaning and only accompanied the breath of the human voice *(flatus vocis)* and had their single entitlement and their single place of origin in the conceptual strength of human reason *(in mente)*. This constructive concept of the power of the human spirit presupposed subjectivity, and therefore a certain sovereignty over time, within a rational-linear understanding of time. The understanding of time developed parallel to the philosophical discussion. We now want take a closer look.

Thomas of Aquinas had absorbed the theory of time of Aristotle for the theology and philosophy of the high Middle Ages. Already we have remarked on the replacement of the Platonic conception of time by the Aristotelian, as Aristotle connects the understanding of time with motion: *tempus est numberus motus secundum prius et posterius.* Time is the measurement of motion with reference to the preceding and the following. This definition raised three problems, however, which were intensively discussed by the generation after Aquinas. Especially at beginning of the fourteenth century, numerous treatises by the aforementioned authors appeared on the problem of time. In this discussion, we will abstain from presenting the numerous variations of solutions. We will restrict ourselves to the formulation of the problems and to the solution of Ockham, which was the most significant and which had an impact on history.

The first problem was that the Aristotelian definition left open the question as to whether time was a subjective or an objective phenomenon. Both solutions were represented, together with a middle solution *(partim . . . partim)*, and they would correspond to our thoughts about endogenous and exogenous time.

The second problem referred to the construction of the continuum of time: Did it consist of points (an infinitely divisible continuum) or of discrete units (quantification), time quanta, so to speak? Here also, there were camps supporting both solutions.

Most significant, however, was the problem of the unity of time. Since time was the measurement of motion, the question was: Which motion was meant? Aristotle had distinguished four types of motion: change of place, coming into being, going out of being, and change of quality. Which movement was the standard for the measurement of time?

At first, we want to consider this issue on the basis of the external and objective astronomic realities in the framework of the Ptolemaic-Aristotelian model of the world, which Ockham also accepted. According to this astronomic model, the earth was a sphere and rested in the center of the world. Turning around it were seven crystalline shells (spheres), in which the seven planets (known at that time) were fastened. The sun and moon were counted as planets. The crystal shell nearest the earth carried the moon, followed by the shells of Mercury, Venus, the sun, Mars, Jupiter, and Saturn. These seven spheres were enclosed by the one that turned with the fixed stars of heaven. It was the eighth sphere. That was why the ninth crystal shell rotated without stars. Beyond the ninth heaven was God, who was beyond space and time, and who rested together with the blessed and angels in heaven *(coelum emphyreum)*. In spite of being the "unmoved mover," God initiated the motion of the different heavenly spheres and upheld them. The pure crystal sphere of the ninth heaven played a special intercessory role between the motionless God and the moving spheres. These spheres were regarded as "the first mobiles" *(primum mobile)*, whose nonstop, fastest movement was attributed to the intercessory authorities, such as angels, spirits, and intelligence that informed the lower spheres. The lowermost sphere of the moon marked an important boundary between the completely regular movements beyond the moon *(supralunar)* and the imperfect movements that were completely irregular on this side of the moon *(sublunar)*, on the earth. An appraisal of value was connected with the hierarchy of movements. The irregular earthly movements enjoyed the lowest prestige, and the movement of the ninth sphere *(primum mobile)* was the highest. Consequently, this hierarchy of movement was connected to a corresponding value of the quality of time.

The questions now are: In view of these different forms of movement and qualities of time, can we still speak of the *unity of time?* Were there not incommensurable times, each of which was completely different from the others? Could one even identify the movements of the lower spheres with the course of time? Ockham dedicated a whole *Quaestio (Quaestiones in libros physicorum Aristotelis)* to this problem. It was characteristic of Ockham's nominalistic attitude that he treated the ontological question of the unity of time as a theoretical one, and he asked about a standard for comparison and about the measurability of time. As a standard for comparison for the measurability of time, Ockham allowed every regular movement, that is, supralunar movement. For example, the regular movement of the sun had been suitable for measuring time, and it fixed a standard for comparison for other variable movements and times (de Ockham 1987, 357). In this way, Ockham used perfect supralunar movement as a standard for the imperfect earthly sublunar movements. This action was remarkable. Ockham carried out a certain devaluation, indeed a desacralizing of the entire supralunar time, in that he made it operational and subordinated it to rational action. This tendency of fetching time from heaven to earth was strengthened when Ockham deemed it to be legitimate to reproduce the supralunar unity of time by sublunar, regular, that is, completely mechanical movements on the earth, by means of a clock (de Ockham 1987, 510-11):

> Question 43: Whether any more inferior movement is time. . . . To this question I say first that a more inferior movement, by knowledge of which we can get to the perception of any heavenly movement unknown to us, can be called time. This conclusion is brought to light daily by many experiences: because we measure the movement of the sun and our actions by the clock, it becomes clear to us that this movement is uniform and regular, because the clock is known to show how far the sun has gone through its orbit, even if the sun is continuously under a cloud.

Ockham completely tore down the wall between the imperfect sublunar movement and time and the supralunar movement and time when he defined reciprocally the movement of the sun and the movement the clock (de Ockham 1987, 511):

> By the movement of the sun, the movement of the clock is measured and directed in such a way that, as soon as the first movable (*primum mobile,* i.e. the ninth sphere, concerning which see the comment above) completes its daily movement, the clock will accomplish its circle; and so it measures the movement of the clock by its daily movement; and af-

ter the movement of the clock is known, it can later measure other movements, e.g. daily movement and the movement of the sun, and to be consistent everyone can call both movements time with reference to the other.

The result was that Ockham did not formulate the ontological question about the unity of time in the context of the Aristotelian-Ptolemaic philosophy as a theoretical question of perception, that is, as a question of the frame of reference and the measurability of time. In this way he desacralized time and also secularized it. Through its connection with regular courses, as in a clock, he brought in the concept of linear time. One could interpret the clock as an algorithm for the manufacture of discrete, regular units of time, and the result of their summation was a *continuum of linear time*. The unity of time had been lost because of the numerous irregular and strict cycles that were deviant movements in nature, as well as the qualitative separation that never really existed between sublunar and supralunar time. The unity of time was restored in an artificial way by an artifact of culture, which now reformed the time of nature. The artificial, regular movement of the clock casts out the rhythm of natural time, and it received its philosophical justification through Ockham. In this way he replaced the ontological unity of natural time with the methodical unity of cultural time.

As far as the question of the unity the time in outer, objective, astronomic space, Ockham's more exact thinking on the question of the relationship opened time and movement from the point of view of the measurability and the subjective character of time. In "Quaestio 40" of *Quaestiones in libros physicorum Aristotelis,* Ockham asks the question whether or not time and movement are identical (de Ockham 1987, 501). The answer corresponded to his nominalistic attitude. In something of a softened tone of voice, he identified time and movement in the ontological sense. Time originated as an accompanying phenomenon of movement, so to speak, and time had no existence without movement, no autonomous being in itself (de Ockham 1987, 504). In this respect, one could not recognize time in itself, and it did not even make any sense to ask what time is. In nominalistic terminology it was only a connotative concept. Concerning the relationship of time and movement, its objective measurability was decided nominalistically by movement, such as by clocks, but subjectively the soul determined that time was measured by movement: "because time is movement, by which the soul knows about the measurement of other movement. And therefore it is impossible that time is time, if not through the soul" (de Ockham 1987, 504).

Ockham's theory of time was significant in a twofold sense. On the one hand, it marked an extremely subjective standard of living, in that time became a measurement of the soul, of the consciousness of outer movements. On the other hand, it bound time by its emphasis on the measurability on regular courses of objective realities, in that it interpreted time as an accompanying phenomenon of movement. The peculiarity of this double, objective and subjective, anchoring of time consisted in the fact that the consciousness of time was experienced as a continuous stimulus by its position in between subjective and objective time. Thus, the rational-linear concept of time moved onto the subjective level. One can rightly wonder, however, whether this intercessory work between subject and object was asking too much of consciousness. Historically, the philosophical discussion and formation of practical life fell back again and again into the unfruitful alternatives of the "subjective" and "objective." For the subsequent time, Ockham's binding of time to the strength of consciousness that constituted time — in this respect Ockham was extremely modern — was crucial, as was its banishment into the homogeneity and lack of quality of mechanical operation.

Once time was God's. Now, it was only a pattern in a machine and a silhouette of the human intellect. In the subjectivization and mechanization of time,[17] Ockham secularized God's time.

c. *Time and the Invention of the Clock*   Besides the rational change, an innovation has persistently affected the structuring of the experience of time and the formation of time: the invention of the clock. It owed its development to two decisive bearers of culture in the late Middle Ages, the cloisters and the cities that were growing economically and in population. In antiquity, the division of days was dependent on the length of daylight. That is, a unit of time was shorter in the winter than in the summer. In the cloisters, however, detachment from the ancient division of the time of day was accomplished with their times for Christian prayers in days of 12 hours and nights of 12 hours spread equally through the seven days. The

17. A detailed explanation of Ockham's understanding of time is provided by Maier 1955, 65-137; Perler 1988, 193-227; Shapiro 1957.

hearers created the necessary constraints on themselves for the homogeneous, linear management of time, which was independent of the length of daylight (Bilfinger 1969; Dohrn-van Rossum 1989, 49-60).

The modern mechanical clock had been invented with weight-driven gearing and an adjustable escapement (the tick of the clock!) in the late thirteenth century, presumably in the cloisters. The exact place of the invention of the escapement has not been located, but already a generation later in 1328 Richard of Wallingford (ca. 1291-1336) (North 1976), abbot of the South English cloister of St. Alban's, created a clock that was very much admired at that time. It displayed not only the time but also the daily path of the sun, the moon, and the most important stars.

After the mechanical clock had been invented in the cloister, it found its place on town halls and churches for public display of the hour. There is firm evidence for the first church clocks in England: Exeter, 1284; St. Paul's Cathedral, London, 1286; and Merton College, Oxford, 1288. On the continent, much more complicated copies were created in the following generation: Milan, 1309, and Padua, 1344. Germany followed with the clock for the Strassburg cathedral in 1352; France with that of Avignon in 1353, and so forth.

The invention of the striking mechanism was connected with the modern way of counting the hours and the clock's implementation in public life. The bases were laid for the organization of a rational-linear time for social life, such as schedules for school instruction, meeting times of public committees, and finally counting the hours related to remuneration for work. To appreciate the meaning of this rational-linear time, it must be understood that it was human time. That is, that it was disconnected not only from the steering rhythms of nature but also from of the church year and holy history. The detachment of time from its natural rhythm (daylight) made it quantifiable, calculable, and open to the public. From then on, time could be organized into abstract units. For the organization of work it meant that work could now be organized quantifiably in time. Earlier, the organization of work was modeled by its yield. There was much work in the summer because of harvest, and little work in the winter, which brought more opportunity for leisure. Now work could be organized independently of daylight and the seasons because it could be independently quantified. This process of disconnecting linear time from natural time and establishing an artificial social time for culture has been increasingly strengthened in modern times, so that night can be turned into day, and winter can be turned into summer — at least under artificial conditions. The detachment of time from its qualitative, liturgical context was added to the emptying of meaningful holy history. Henceforth, individuals had to manage the meaning of their own time. Briefly, rational-linear time was removed from nature, from quality, and from the sacred, and thereby it became meaningless.

The relationship between God and time shifted with the invention of the clock. The great theologian and mathematician Nikolaus Oresme, in the year 1370 (Wendorff 1985, 144), noted this change of relationship by comparing the world to a big clock that was created and kept running by God. Yet the actual realization of quantifying time, which was now possible in principle, did not begin until the end of the fifteenth century with the regulation of practical life. Even a uniform numbering of the years *post Christum natum* had in no way been asserted in Europe. There is no talk yet about the modern experience of the pressure of time at work. We now come to the economic aspect of the organization of time.

d. Time and Economics   The openness to the future, connected with the concept of rational-linear time, made time into an economic factor. In principle, efficient action could now achieve future profit on an investment. Credit for the funding of a future undertaking was the financial form of trust in business. Its repayment was safeguarded in interest. This type of management marked the transition from the agrarian-rural subsistence economy to the urban capital economy in the Middle Ages (Rinderspacher 1989, 91-104). In the subsistence economy humans were only interested in self-preservation. In the capitalistic economy humans developed an interest in actively shaping economic policy. In a capitalistic economy, the economy was tied to action, which implied that time was at their disposal. Here there were theological misgivings in the Middle Ages because, according to the opinion of the church, God was the Lord over time, not humans. Therefore, it was only indirectly over the rate of interest that humans could claim to reign over time.

The usurers taking interest were quintessential sinners, and they had to be prepared to fall into eternal damnation in hell.[18] The Lateran Coun-

18. The attitude of the Bible toward interest is not clear. While the Old Testament forbids the taking of interest (Lev. 25:35-37), this command is not repeated in the New Testament. On the contrary, the parable of entrusted money (Luke 19:11-26) can be understood

cil of 1179 decided to exclude usurers from the sacraments and to deny them a Christian burial. This regulation was confirmed at the Council of Lyon 1271. Accordingly, the public reputation of usurers was quite low. In the fourteenth century, the Franciscan monk and preacher of penitence, Bernardino of Siena, commented about usurers: "the death of an usurer resembles the death of a pig, because his whole life prepares the pig only for loss . . . if however it is dead, all are glad" (Origo 1989). Earlier the church father Augustine (354-430) had considerable reservations against the taking of interest, for it increased a fortune, he argued, because interest obviously grew in time, and therefore without human action. Time belonged to God and not the people, and therefore it was not allowed for them to appropriate it for themselves by means of lending money. In the Middle Ages, people turned to the Jews for disreputable transactions of money. The practice skipped over the reservations of the church.

At this point we must also consider the theological dispute over poverty that was kindled by the Franciscans in the church and was important for the development of economic thinking historically and intellectually. Canonized as a saint in 1228, Francis of Assisi (1182-1226), in a rule approved by the pope, had prohibited his brothers from so much as touching money. For absolutely necessary payments, they used a small rake, and they did so to the mockery of the people. In principle, it was a question whether in the discipleship of Jesus the church could be rich, or whether it ought to distribute its fortune to the poor. Pedantic theological opinions in the great universities in Europe were involved in deciding which attitude Jesus himself had toward money. Finally, a decision was brought about by the power of Pope John XXII in the year 1323. He gave his opinion that the Franciscans were heretics in saying that Jesus had lived as a beggar. The church thus pointed the way toward unrestricted economic thinking and action. The first test case for this new economic thinking was the church itself, as John XXII sought to restore, among other things, the ruined church finances by means of double-entry bookkeeping, which had been invented in the Upper Italian cities.

Since he was a Franciscan, it was remarkable that Peter Johannis Olivi —

---

precisely as encouragement for the formation of business capital. However, the Lord's words in Luke 6:34-35 [NRSV] ". . . and lend, expecting nothing in return" could be interpreted as hostile to it. That is how it was understood in the Middle Ages, and it served as exegetical evidence for the so-called canonical prohibition of the taking of interest. However, Thomas of Aquinas added certain exceptions (*Summa Theologia* 11.2 *questio* 78, *art.* 2-5).

already known to us for his tracts concerning time — was an original thinker in economics at the end of the thirteenth century. He weakened the theological argument against usury and therefore the instruction about time, which was God's. Olivi introduced a pregnant distinction between God's time, which was not available to humans, and the time of creation, which was the time of humans (Bettini 1953, 148-87). With this small but long-range distinction Olivi snatched, so to speak, from God the sole control over time and changed it to a configurable object like the rest of creation.

On the basis of created time, humans could now charge interest, provided it corresponded to an equivalent work, without making themselves guilty of sacrilegious arrogance by going past their natural limits (Kirshner and Prete 1984, 233-86). This distinction in the concept of time brought with it a corresponding distinction in the concept of money. Olivi distinguished between money and capital. It was his intellectual and historical achievement to have first coined the concept of capital. Capital was allowed to "work" in the configurable medium of created time as a corresponding equivalent of work, such as in the form of credit that bore interest: "If money or property in a secure business is positioned for a likely profit for its owner, it has not only the simple quality of money or goods, but in addition a certain seed-like quality for profit *(quandam seminalem rationem lucrosi)* which we generally call capital *(capitale)*."[19]

On the basis of this distinction, near the end of the thirteenth century, Olivi could recommend to the city of Florence that it finance with credit its expansion of the city nearly sixfold. On the one hand, Olivi advocated the accumulation of capital on the objective side. On the other hand, subjectively and psychologically, Olivi was an opponent of those who supported the strict ideals of poverty, or a consumeristic or even hedonistic attitude toward money and property. The required denial of urges, and the need for delay, build the necessary psychological tension for thinking of economic investing in time, and they were the prerequisites of entering into the accumulation of capital. Here we can see a certain prefiguring of Max Weber, who so brilliantly analyzed the phenomenon of the "inner-worldly asceticism," which helped the capitalistic spirit gain a historical breakthrough in Protestant countries.

Piquant proof was found in the interplay between the theoretical con-

19. Cited in *Die Zeit* 9, no. 21 (February 1997): 43; more exactly investigated in Olivi 1980, 1990; Kirshner and Prete 1984.

solidation of the Franciscans and the power of Pope John XXII. It smoothed the way for capitalistic thinking, and also for thinking of economics in terms of time. Then in the fourteenth century, mainly in the cities of upper Italy, trade over longer distances found favor for the building of money and capital. A developed world of banking was created, with a corresponding money and capital market, which drew in an immense concentration of capital. Kings and popes were the preferred, and usually tardy, borrowers. This problem of their insolvency and unwillingness to make payments was not rarely solved by expropriation of the creditors (Beckerath 1956, 5:480-81).

Through the first wave of the plague (1348-52), this situation was partly responsible for the first great depression, and also for the greater scope of the phenomenon of inflation. Already in 1355, some decades after the establishment of the capitalistic economy and three years after the plague, Nicole Oresme, the philosopher of nature and later bishop of Lisieux, was the first to be concerned with the theory of inflation (Oresme 1503).

Despite its theoretical and practical contribution, the attitude of the clergy in this phase of early capitalism was absolutely contradictory in regard to the enormous accumulation of capital. The Franciscan emphasis on poverty still continued to have an effect. Its preachers of penance railed against the pursuit of money and against profiteering and cheating, and they upheld the ideal of correct measurement, as well as poverty. Again and again great preachers from this Order engaged in polemics against the excesses of the widening capitalistic economy. The preacher of repentance, John the Baptist, was the best known biblical character, and he was the one most often publicly displayed at that time. From the fourteenth to the fifteenth century, the popular Franciscan Bernardino of Siena worked (1380-1444).[20] In the fifteenth century the Dominican Girolamo Savonarola (1452-98) at times changed Florence into a dictatorial theocracy in which the accumulated riches of the merchants were burned publicly as a work of penance.

Only in the Reformation, specifically with the Reformed faith, did the thought come through that humans could assure themselves of God's

---

20. Origo 1989, 68-85; just as with his earlier fellow member of the order, Petrus Johannis Olivi, it also applies to Bernardino as a far-seeing economist (cf. Roover and Schumpeter 1957, 125-27).

blessing by economic success, and thereby assure their personal predestination to blessedness. The conception was theologically legitimized that humans were called to be responsible by God *(vocatio)* and were loaned time by God as responsible economic subjects. Max Weber (1864-1920) impressively presented the connection between Reformed belief in election and economic activity. If the medieval subsistence society was distinguished by a lack of goods and an excess of time, then in modern times the relationship has been reversed: many goods, little time. Economics had changed *time* to the *management of time itself:* time was a scarce commodity, time was money.

e. *Time and the Plague*   The years 1348-52 marked a turning point in European history. In this short time, there are estimates that the plague, the black death, killed between 30 and 50 percent of the European population. Whole regions were depopulated, and public life collapsed. In the following centuries the plague continued to flicker in individual regions of Europe again and again, in larger and smaller measures until approximately 1720. This excruciating form of dying changed the attitude of the European people toward their own deaths. People stricken by the plague could no longer prepare themselves inwardly for death through the art of the *Ars moriendi* practiced *bis dato* — they died a premature death. They had to do so without the community or their relatives; they were not worthy; they were not supported by the sacraments of the church, in order to reach from this side to the hereafter. Death by the plague meant the end of all kinds of life and death that had been valid up until then. Death by the plague, because it was an untimely death, was no longer the passageway to eternal life; it was the enemy of humans.

Marianne Gronemeyer explained the thrust toward individualization from the traumatic experience of the plague: "A global order offers the individual an unquestionable place in a structure of meaning, and when it collapses, a thrust toward individualization arises" (Gronemeyer 1993, 20). The plague was a quintessential breakdown of all outer and inner order that gave meaning. The inner and outer order also referred to time: "If humans could dare to marry, to pursue the subjugation of time to their reign, God would have to be stripped of power as the Lord of time. Similarly, the mastery of nature happened first, as nature, instead of being God's gift, became powerful in itself and hostile" (Gronemeyer 1993, 76).

Subjectivization and individualization made God's reign over time outmoded. The humans who were dying and the ones who were surviving had been robbed of all their outer and inner supports, and they were thrown back on themselves completely. All the consequences of the epidemic of the plague favored this individualization. The bonds of relationships were relaxed because of the fear of infection, and humans became distrustful of each other and isolated. Since death could no longer be assumed to be the passageway to eternal salvation, life for humans was completely directed to this life. Two options were open. Humans could either be desperate over the inevitability of their fate or *carpe diem,* enjoying worldly pleasures in all its nuances, from primitive gluttony to refined aesthetic charm. Boccaccio (1313-75) portrayed this subtle, artistic worldly pleasure in his *Il Decamerone.*

There was, however, a third option. Examining the plague, Marianne Gronemeyer noted that surviving humans took on the fight with death, in that they made the extension of earthly life the highest goal of their efforts. Through planning humans now experienced sovereignty over time, and they wrested from God the rhythms of nature, order, and interpretation. This planned time was our rational-linear time. Maybe it was no accident that turning time into an economic commodity by mechanical clocks coincided with the beginning of the plague. The high rate of mortality reduced the number of viable people considerably. Work suddenly became a precious asset. The depression it caused in Europe required more efficient relations with only the precious economic factors of work and time (Sarton 1953, 1653). In this context, it was after the invention of the clock in the cloisters that clocks could begin to form social time in the public towers. In the first year of the plague, 1348, the Italian Giovanni of de Dondi began to construct a highly complicated clock, which was set up sixteen years later in the market place of Padua. Starting in 1350, there was a boom in Europe for public clocks that would strike the hour.

f. Time and Music Music can be understood as the art of the formation of time, and thus we expect that our thesis would be substantiated here with the strengthened individuality of the rational-linear consciousness of time. Musicologists are united in saying that the fourteenth century produced the first individually inventive musicians and composers: "For the first time, free artistry emerges in polyphonic music from an inventive personality. The great form of the isorhythmic motet is autonomous, and it does not fit in the liturgy or some other social order" (Blume 1989, 714).

This new type of music appeared in France and Italy, the two leading countries of the late Middle Ages. In Italy the art forms of the madrigal and the ballad (music of the 1300s) were new and connected with time, as was the isorhythmic motet in France (Ars Nova). The Ars Nova was so named by the chief representative of this work in musical theory, Philipp de Vitry (1291-1361), in the year 1320. In the twentieth century Hugo Riemann coined the term as the name of a musical epoch, which developed from the liturgical tradition of the Notre Dame School of the thirteenth century, which Philipp de Vitry had called Ars Antiqua. What was the newness of this Ars Nova, especially in regard to the rational-linear structure of time, which we have postulated as new?

Already in the Ars Antiqua, which was connected with liturgical polyphonic music of the thirteenth century in the cathedral of Notre Dame, intramusical tendencies had developed. These tendencies required more economic and more highly structured relationships in the course of time as a medium of music. Up to that time, sound was prevalent in the motet of two voices, and the musical elements had been created for spatial effects. Now, time and its organization increasingly emerged in the foreground for compositional and technical reasons. In the epoch of the Ars Antiqua, about 1250-1320, the crucial element of musical style was still the combination of sound and space: "Musical sound and space, spatial sound and sound of space, color and light yield a unity of complete beauty. The unity that is found is not only optical and acoustic, it is also a mental unity. It corresponds to the theological-philosophical unity of the philosophy of the epoch, which is named the scholastic epoch" (Nestler 1975, 104-5). In Gothic cathedrals, the timing of reverberations that lasted several seconds for low frequencies reinforced the adjustment of sound in music. In the epoch of the Ars Nova, around 1320, the arrangement of music in sound and space was replaced with music arranged by movement and time. This connection of time and movement was made the subject of discussion *expressis verbis* in the first work on musical theory in the Ars Nova, the *Ars novas Musicae*, by Johannes de Muris, in the year 1319: "Therefore, the voice must necessarily be measured by time. It is particularly the time of the measurement of movement. But here time is the measure of the voice that is produced by ongoing movement: The same however is designated as a definition of time and unity" (Gerbert 1963, 292).

This new alignment of music had already shifted during the time of the Ars Antiqua. Formal musical problems contributed to the shift that finally led to introducing time ever more strongly into consciousness as a constitutive element of music. To this shift belonged especially the polyphony that arose in four voices, the further working out of counterpoint, as well as an increasingly complex polyphony and rhythm. In polyphony it was true that the separation and coming back together of the voices was coordinated in time. Stimulation for this development led onward to the development of interval and harmony. Also, rhythm puts its accent on time.

Finally, the values of notes had to be designated unequivocally. Up to that time the Ars Antiqua had justified religiously the complete *(perfectum)* three-sided division of the value of notes by referring to the doctrine of the Trinity. The Ars Nova replaced it by the worldly imperfect *(imperfectum)* two-sided value of notes, and it has continued up to the present time. The tempo of music could increase because the reference point for the timing of music changed from the breve to the shorter semibreve. All these aspects required a precise regulation of time, and the result of these technical necessities of composition was the replacement of the modal rhythm of the Ars Antiqua by the measured rhythm of the Ars Nova. The established note values were now determined by measured notation.

We have thus arrived at the key concept of the Ars Nova, which Philipp de Vitry had analyzed in technical detail in his book *Ars novas* in 1320. Time could be understood in principle as a continuum[21] that could be resolved into equal units. That is, time had to be countable, and a system had to be created for the order of time and number. Actually, the reference point for the musical counting of time appeared in the higher rhythmic complexity of the semibreve. The availability of the semibreve, besides the breve and long, now increased the possibilities for combining different lengths of the values of notes. However, a new phenomenon appeared in the hearing of music. As possibilities for composing melodies were increased by the different values of the length of notes, these new possibilities put a demand on the memory and the expectation of the conscious-

21. Surely it is no accident that the concept of the divisibility of time is supported in natural philosophy by the contemporaries of Philipp de Vitry and Johannes de Muris. In their study of Aristotle, they were trying philosophically and mathematically to grasp more exactly the concept of a continuum. In principle, there is a question as to whether the continuum is divisible in any way, or whether it consists of discrete units. Here we have to do primarily with William of Ockham, Johannes Buridan (1300-1358) and his school (Nikolaus von Oresme [1325-82], Albert von Sachsen, and Marsilius von Inghen), as well as Heinrich von Harclay (cf. Maier 1949, 155-215; Gericke 1993, 137-63).

ness of hearing, in order to keep present in the consciousness a longer and more complex melody as a whole. This was only possible by presupposing a subjectively rational-linear time.

Interestingly, the theoretical consolidation of the musical opinion of time was achieved by the mathematician and theoretician of music Johannes de Muris in his writing *Ars novas Musicae* in 1319. He was an intellectual comrade-in-arms with Philipp de Vitry in the Ars Nova, and his book appeared one year before de Vitry's book. In the second part of this treatise, the *musica practica*, he avoided the management of time.[22] At first, he used mystic-anagogic theology in the tradition of Dionysus the Areopagite (ca. fifth century A.D.), which especially dominated the theology of the Abbot Suger of St. Denis in Paris. He placed time in the context of the individual steps that summarized the being of the Trinity. The Trinity corresponded to the threefold division of the values of notes. The completeness of the Trinity opened up in an image, which was in accord with the Augustinian teachings about the *vestigia trinitatis*. Time, the stable wholeness of the Trinity, and its image of the three-sided division of notes were related to one another. "At every time, predecessors of the measured voice have rationally attributed a certain measure of perfection. They determined that time is such that it can be divided by three, and they think that perfection is found in threeness. Therefore, they put this perfect time as the measurement of any chant, knowing that the imperfect cannot be found in art. On the contrary, certain modern people think they have found what is missing" (Gerbert 1963, 292).

This perfection rested in depicting threeness, and Johannes de Muris replaced it with imperfect twoness. However, the twoness *(binarius)* was based more on movement, more diverse combinations of lengths of tones, and rhythms. It made room for musical development: "The opposite of it [i.e., the threeness, see comment above] is the case, if the binary is separated from it [the threeness]; then it [the threeness] remains behind as incomplete. But also the binary is bad" (Gerbert 1963, 293).

Johannes de Muris admitted this earthly imperfection, and he created the dimension of *temporal development* in music. Time as a musical dimension was a product of earthly imperfection. Finally, after detailed dis-

---

22. "It brings as a kernel a mathematically based division of time and a system of rhythm that is based on the distinction of the four *gradus*. It corresponds to the scale of the Ars Nova" (Blume 1989, 7:109).

cussion about the theologically motivated three-sided values of notes and the more technically musically inclined two-sided values of notes, he came to nine conclusions. The last involved the nature of time. He wrote: "Time can be split into many equal parts, which are brought to light. The whole continuum is divisible into many parts in any proportion, as in two, three, four, etc. Time is that kind of continuum. Consequently, it can be divided at will into equal parts" (Gerbert 1963, 300).

He defined time as a linear continuum that was divisible at will. Besides the theologically perfect three-sided notation, Johannes de Muris allowed the earthly two-sided imperfect notation. At the same time he carried out a kind of desacralizing of time. He subordinated time to rational calculation as a continuum without qualities.

The measured notation based on this conception of time became generally accepted from 1320 with Philipp de Vitry's book *Ars nova*. With the *Ars nova*, music left its linkage with the church and its liturgy and became free art. Furthermore, church music in the south of France had already become "free" as a courtly function, from around 1100 to 1350. Champions of this new musical movement in France were Philipp de Vitry (1291-1361), Johannes de Muris (ca. 1290-1351), Guillaume de Machaut (1300-1377), and, in Italy, mainly in the transposition of time, Francesco Landini (1324-97). These musicians also participated in the intellectual performances of other sciences and arts. De Vitry was friendly with Petrarch, as were also Johannes de Muris and Landini. Incidentally, in his lifetime Petrarch was insistent on punctuality, and he was notorious for it. Landini was a supporter of nominalism, and he even wrote a long poem in praise of William of Ockham.

It is not remarkable that opponents of this plan of musical work called it a revolutionary transformation. The musicians' guild rejected the Ars Nova, as shown in the chief work that summarized and concluded the Ars Antiqua, the *Speculum Musicae* (1324-25) by Jacob of Lüttich (1260-1330). The church viewed this secularization and individualization of music with suspicion. Thus, Pope John XXII (reigned in Avignon 1316-34) prohibited the Ars Nova in the church in the decree *Docta Sanctorum* 1324-25. This interesting document was directed primarily toward active individuality, and the pope was stepping in as administrator of the truth of the church, which was beyond personal comprehension:

But several supporters of a new school want eagerly to introduce measured tempo and new notes, rather than producing their own [chants] as the ancients recite; in semibreve and minims will decency be sung in the church, which would be marred by the values of small notes.

Because they dismember the liturgical melodies by Hoqueti; by multiple voices they neglect the sentences; vulgar linguistic mincing and moth-like fluttering — all of this shows that they think little of the foundations (the liturgical melodies) of the antiphonies and the graduals: They ignore the One to whom they erect their compositions.

They do not pay attention to keys and they do not distinguish them. Rather they bring them into disorder, since, according to the quantity, the [small] music obscures the natural ascents and the proper downward movements of the *cantus planus,* in which the keys are mutually different.

They hurry, and they do not rest; they intoxicate the ears, instead of calming them; they rush about without rest — they disturb devotion instead of evoking it. (Eggebrecht 1991, 220-21)

However, the decree was not generally accepted. At the end of the century, Ars Nova provided music in the church.

On the one hand, we have shown in this chapter that the rational-linear understanding of time was tied to an increased self-esteem of individuals. On the other hand, we have illustrated through examples that this new perception of time was generally accepted in all important areas of life. The increased individuality and subjectivity led to a certain sovereignty over time, so that henceforth time could appear to humans as a pliable medium, and it could even appear as an object. With help of the clock, humans could measure and divide time; in economics they could make it work; they could accentuate time rhythmically in music. In all areas of life, the fourteenth century made time linear and began to provide it with dynamics. The following generations worked on sharpening the concept of linear time, and they advanced their subjectivization of human experience and its objectification in practical concepts. The philosophy of René Descartes represented a crucial turning point in this sense.

Descartes (1596-1650)

For René Descartes, the question of certainty was the fundamental problem of his philosophical life. As a pupil of a Jesuit course of lectures famous in his time, he was entrusted with scholastic philosophy, although no certainty could be found in their deductions. So, he looked for some unambiguous basis for all philosophy. He developed a method of doubt in the hope of finding something that could withstand doubt. After a long spell of spiritual exercises in this method, he found something about which there could no longer be doubt. In the center of human consciousness he found the following: "I think, therefore I am; *cogito, ergo sum.*" That happened in the year 1619, while he served in camp for the winter at Neuburg as an officer in the Bavarian army.

Since this consciousness of the "I think" was a temporal consciousness (I sense continuous changes in consciousness), time was in human subjectivity and not in the processes of the external world. Therefore, Descartes sharply dissociated himself from the Aristotelian concept of time, which determined time as a "something" in the external movement of natural bodies. He wrote in his "Principles" (1.57.27, in Link 1978, 219):

> If we distinguish time *(tempus)* from duration *(duratio)*, and say it is a measure of movement, then this is merely a condition of thinking *(modus cogitandi);* for we actually notice no duration in movement that is different from duration in motionless things. We can illustrate it this way: if there are two bodies — one fast, and the other slow — that move for one hour, we count no more time in one than in the other, although the quantity of movement in one is much greater. In order to measure the duration of all things, we compare it with the duration of the largest and most even movement, which is derived from years and days, and we call this duration "time" *(tempus)*. Taken in the most general sense, therefore, appending duration is nothing other than a condition for thinking of it.

After Descartes, the temporality of the consciousness had a transcendent cause: it was performed by God as the most complete and freest essence, from which nothing greater could be intended. Christian Link expounds on it (1978, 240):

> This duration that goes along with the movement of thinking *(modus cogitationis)* "is" nothing other than the consciousness of time which is loaned by God for existence. At the same time, if this consciousness is the subject of our experience of time, then the temporal essence of duration measured in it is comprehended as coming from God. So that proves that the same duration of God is without beginning and without

change, in which all created things enter into their temporal existence, according to the basis of the Cartesian concept of time.

That is to say, time could be seen as no more than the essence of its natural reality. Its horizon was opened by the duration of God.

The "break" between the time of the consciousness that presents it and the modes of time that are structurally divided up into present, past and future becomes an irreplaceable part of everyday experience. However not only that: the changed horizon of time justifies the necessity of projecting time into the area of timeless truth. Only with this condition, can one grasp it "clearly and distinctly" in the consciousness. In this way the changed understanding of time was required by Descartes' "mathematical" access to the world. (Link 1978, 244)

In this way modern natural science began to describe and to understand all the events in nature *mathematically*, and also to anticipate events in nature. With his system of "Cartesian coordinates," Descartes contributed to the development of analytic geometry. In the categories of analytic geometry, time was understood as a horizontal line as one axis of a coordinate system, and so it was pressed into a scheme that became constitutive for the concept of time in classical mechanics. Descartes therefore worked with an understanding of linear time. As with Ockham, he transferred it completely into subjectivity, but he departed from it, however, in that time was no longer an act of the intellect to which movement was appended, but rather was only connected with acts of consciousness. Movement as the constituent element of time was discontinued completely, and so time also was made fully linear. This linear continuum of time was a first-rate description of movement in nature. It was now only a small step to the concept of linear time in science.

Newton (1643-1727)

Early in his study in Cambridge, Isaac Newton argued with Aristotle and Descartes. Natural scientists accepted Descartes's procedural methods, with deductions from his physics. The general intellectual climate left behind Aristotelian philosophy, whose scholasticism appeared to be scientifically out-of-date and not acceptable religiously, especially by the biblicism of the Puritans. So, it is not surprising that movement no longer appeared to be a constitutive factor in Newton's opinion of time. Concerning Descartes's connection of time with consciousness, we find out nothing further. Descartes's understanding of time as a parameter in the coordinate system became all the more significant. Through his conception of time, Newton's teacher Isaac Barrow (1630-77) influenced Newton's understanding of time:

> Time does not mark an actual existence but a decided capacity or possibility for the continuity of existence. It is exactly the same as area is a capacity for the length lying on it. As far as its absolute and actual nature is concerned, time presupposes movement just as little as it presupposes rest. Whether the things move or are at rest, whether we sleep or wake, time continues its regular pace on its way. Time presupposes movement in order to be measurable; without movement we do not perceive the course of time. We apparently must view time as a constantly flowing river. (Whitrow 1973, 114)

Barrow completed the connection of (absolute) time with movement. Movement was not inherent in time; it was only an external method for measuring it. It was especially remarkable that Barrow had already suggested the concept of the absoluteness of time, which received its classical formulation from Newton.

This definition represented the conclusion of all previous idealizations of the understanding of linear time and a victory of abstraction over the experience of natural time. The relative time of the experience of natural time was devalued in favor of the *unity and universality* of absolute time. Henceforth it was basically a coordinate and a universal reference underneath all variable movement. This idealization and homogenization of time had the advantage that it was divisible at will mathematically in the form of the differential calculus of Newton and Leibniz, so that movement could be described precisely. The separation of time from the (Aristotelian) concept of movement was imperative. Another idealization also proved to be fertile. As we will see in more detail below, the mathematical concept of time was symmetrical; that is, direction was eliminated from the past into the future. This idealization in the form of symmetry made possible the formulation of the invariance of time in the laws of nature. We established three characteristics of Newton's concept of linear time: *abso-*

luteness and *universality,* any amount of *divisibility (homogeneity),* and *symmetry.*

Despite the great success of these three idealizations, they soon evoked opposition. Gottfried Wilhelm Leibniz (1646-1716) attacked the idea of an absolute time in favor of a relative understanding of time, and Ernst Mach (1838-1916) made these reservations more specific from a positivist perspective. However, Albert Einstein (1879-1955) dethroned Newton's concept of time. Einstein realized that the speed of light is a constant in all frames of reference and postulated the equivalence of all observers. Also, the relativity theory made clear that the measurement of time depended on movement. A belated victory for Aristotle!

The question must remain open as to whether the idealization of the symmetry of time will be of help in the direction of a universal arrow of time in the future.

Kant (1724-1804)

Newton's doctrine of absolute time was analyzed critically and was altered in the transcendental philosophy of Immanuel Kant. In his *Critique of Pure Reason,* space and time were not understood as objective forms of things; rather, Kant gave reasons why space and time were subjective forms of our human viewpoint. According to Kant, time was a subjective scheme of order, with which we "place before" ourselves the perception of given objects: "Time is the formal condition *a priori* of all appearances."

We humans cannot perceive and experience the world and ourselves other than temporally. With Kant, the essence of time, or, to say it more exactly, the unity of time, was introduced in the tradition of the Greeks' teachings as to eternal identity and presence. Kant said in the chapter "Of the Schematism of Pure Concepts of Reason" (Kant 1990, B 183): "Time does not disappear, but existence disappears in the changes in it. Since it is unchangeable and permanent in itself, time corresponds to the unchangeable in appearance in existence, i.e. substance."

This citation corresponds to another place in which Kant spoke about the "first analogy of experience" (Kant 1990, B 224-25):

> Time, in which all the changes of appearances should be thought, remains and does not change; because it is that in which a definition can be conceived of changes or sameness. Now, time itself cannot be per-

ceived. Consequently, the substratum must be found for perceived objects, i.e. appearances, which introduces time. These appearances can be perceived in all change or sameness by the relation of the appearances in the apprehension. However, the substratum of all reality, i.e. what belongs to the existence of things, the substance, in which everything has existence, can only be intended as a definition.

The essence of time, its unchangeability, therefore is not directly perceptible and conceivable. It appears to us only in the mirror of substance. We recognize the eternal presence in this mirror as the unity of time. The distinctions of the three modes of time (past, present, and future) disappeared in the light of this eternal present. Georg Picht rightly says (1980, 366): "Kant's teaching about time is not the result of an unbridled speculation. He attains it by transcendental reflection on the conditions of the possibility of classical physics. It is a teaching of the hidden implications of this physics. Kant reveals the presuppositions that justify the merging of appearances in time with a parameter defined as a fixed straight line."

Precisely this understanding of time defined the bases of classical physics. It implied the reversibility of natural laws (everything could also run backward) as well as determinism (everything is predetermined by the laws of nature).

However, Kant's philosophy had a greater significance for the social implementation of linear time. With his philosophy of the Enlightenment, the thought of *progress* penetrated into ever wider circles of society. In the Baroque, just as there was a conquest and formation of space by living and forming movement, so also there was a beginning of the conscious formation of time. It is no wonder that Kant became known for his pedantic management of time. Understandably, the concept of progress was joined with the concept of speed and the concept of acceleration. Progress, speed, and even acceleration in all areas of life were the models of the nineteenth and twentieth centuries. With this connection, the concept of linear time seems to be pushed to the limits of its capacity.

The goal of this section was to make plausible the intellectual presuppositions for the birth of the concept of linear time as an expression of individuality and of subjectivity in the fourteenth century. Furthermore, this birth happened with help from the invention of the clock, which spread the concept of linear time widely throughout society, provided for a social commitment, and thus gave rise to a lifestyle. By its continuous philosophical refinement, linear time reached a point where it could become a fundamental idea of natural science (classical physics). Finally, it apparently reached the limits of its capacity through the phenomenon of acceleration.

Discontent with the concept of linear time has appeared again and again, and there have always been people who have tried to reject the linearity of time. That is the concern of mysticism.

Mystic Experience of Time

The mystic realization of timelessness is the rare human experience of the mountaintop. It presupposes and at the same time exceeds the linear understanding of time. This presupposition must be clearly emphasized once again. In everyday understanding, mysticism is often connected with dark, confused enthusiasm and fickle, instinctive exuberance. In science there are tendencies to equate phenomena of mystic experience with experiences examined by psychopathology, specifically schizophrenia. We do not share this view. Rather, in our opinion mysticism is a sublime expression of the highest creative possibility of our humanity.[23] In theological terms, we assign mysticism to creation, not to salvation. On the basis of the experience of mysticism, this view shows us the possibility of acknowledging as meaningful a point of connection for the interreligious dialogue. It rejects, however, a claim that has become fashionable today: the ability to create a type of uniform world religion on the basis of mysticism, or neuro-theology, which presupposes a "God

---

23. A rethinking of science in this direction — making mystic experience a pathology — is primarily indebted to William James (cf. James 1958) and the physician and researcher of mysticism, Carl Albrecht (cf. Albrecht 1951). William James gave a phenomenological description of mystical experiences in his Gifford Lectures from 1901-1902, such as "ineffability, noetic quality, transiency, passivity" (James 1958, 292f.). Unfortunately his description does not mention the experience of timelessness. A broader phenomenological and theological research was undertaken by the physician and theologian, Klaus Thomas (cf. Thomas 1973). On the psychological side, Abraham Maslow comes closest to speaking of mystic experience as the highest form of self-development and of peak experiences ("mountaintop experiences") (Maslow 1989). In more recent times, the meditative process has also been investigated psychologically and physiologically. Engel (1995) offers a survey of the state of the research with detailed overviews of the literature. The recent experimental data found by d'Aquili and Newberg about the decrease and increase of blood flow in mystic experiences confirm our theoretical prediction that the experience of timelessness has a physiological correlate in metabolism (d'Aquili and Newberg 1999, 119).

module" in the human brain (Ramachandran and Blakeslee 1998). God's truth is certainly greater than the mystic experiences of humans. It is remarkable that such experiences usually occur in times of change, and occur with people who could then give their surroundings new intellectual impulses. We bring together some witnesses of mystic experience.

Plotinus (204-70)

In Plotinus the Neoplatonic spirituality of late antiquity came to its consummate and concluding form. Antiquity had passed its zenith, and Christianity, with its specific experience of time, came powerfully to the fore as a historical power. Plotinus was the first philosopher who dedicated an independent treatise to the problem of time and eternity, namely, *Ennead* 3.7, "Concerning Eternity and Time." For him the mystic experience of eternity had the character of an "ecstasy" of flowing time, from which — at least for Plotinus — there was no return. He paraphrased *aion* (eternity) as "lives of the spirit." While he was in the dialectical circle of this concept of *aion*, whose divine, ineffable essence was honored, time was only the wayward son of eternity. It was defined as "lives of the soul" (the world soul was meant), and its "fall" from the unity of time was virtually described mythically and was mourned. With Plato time was created in joy and for enthralling the God of the world. With Plotinus the world was no longer a beaming "image of God," and *chronos* (time) was consequently only a sign of the garbage of the One. Number and divine order no longer played a role in Plotinus's theory of time. His interest in overcoming time by transcendence applied all the more to the eternal present, in the *aion*, which was for the first time identified with God. With Plotinus the irreconcilable difference was perhaps seen most clearly between the biblical view and the Greek view of the transcendence of time: "Blessedness," he said, "was not a pursuit of the future!" (*Ennead* 3.7.4). One who could not assess future perspectives positively did not have a positive assessment of the world. Neoplatonic spirituality was connected with a strong tendency to escapism.

Augustine (354-430)

If Plotinus stood at the end of the heathen antiquity, then, after an agonizing inner struggle about truth as a convert, Augustine was the beginning of the new Christian era in the West.

In the 11th Book of his *Confessions,* in the framework of his famous treatise about time, Augustine tried to unify Neoplatonism and the biblical heritage. He, with Neoplatonic premises, inquired about time and eternity within the horizon of the biblical belief about creation. With the Greeks, he firmly held onto the eternal present of being, and he shared their devaluation of the merging of the modes of time: past, present, and future. With the Bible he held onto the essence of time as noncyclic, contingent, and freely created by the gracious God. In this way, time lost its cosmic anchoring, and it threatened to disappear into the world and meaninglessness. Then, the human soul appeared to him as a way out. With the biblical thought of the image of God, he managed to represent the soul as the place of the manifestation of time and even, to a certain degree, of the establishment of time. In memory *(memoria)*, perception *(contuitus)*, and expectation *(expectatio)*, time was present in the soul. In the simultaneity of re-called past, present, and future, the soul was able to portray widely one piece of the unity of time, of eternity. Precisely here was the image of God. The true unity of all time was in God's eternity, which stood high above time in the "eternal today." The experience of this transcendence into an open future was far removed from Augustine, however: the human soul experienced time as a "race to death" (*City of God* 13.10), as irrevocable transience. Augustine's "pessimism about history" converged with Neoplatonic escapism. The biblical belief in creation apparently was not strong enough to integrate the reality of the world.

Boethius (480-524)

Some years before the birth of Boethius, the Ostrogoth Odoacer had dethroned the last Roman emperor, Romulus. The Western Roman Empire had sunk with it. When Boethius was thirteen years old, the Eastern Roman Emperor Theodoric overthrew Odoacer. As the son of an aristocratic Roman family, Boethius was educated as a Christian. However, his ecclesial Christianity and philosophical-intellectual alignment remained separated from Neoplatonism. In his best-known and probably most influential work, *The Consolation of Philosophy (Consolatio Philosophiae)*, there was no indication of his Christian roots. Nevertheless, he Christianized the Greeks' understanding of eternity and brought to it a logical formula, which had a lasting influence on the Middle Ages.

He wrote *The Consolation of Philosophy* in 524 as a kind of balance of life, reflecting upon the deaths of politicians sentenced to prison. In an interchange with Plato and Aristotle, he defined eternity as "completely simultaneous and fully possession of an unlimited life." Here was brought into focus once again a summary of the mystic experience of time: It was an immersion into the eternal present, a transcendence of mortality in intangible, complete self-possession.

Meister Eckhart (1260-1328)

Eckhart's socialization was completely in the character of ecclesiastical and established Christianity, which a generation earlier had appropriated the intellectual treasures of antiquity. He lived in a time of change, the fourteenth century, which we have already shown to have produced the concept of rational-linear time for the West. Eckhart was at the intellectual height of his day, and his conception of time departed with him. Eckhart was a strong, highly educated personality. He possessed practical managerial qualities, theological acumen, and erudition, and he was a gifted preacher. As leader of the Dominicans, he was entrusted with administrative tasks and the supervision over cloisters throughout Europe, where he traveled — on foot! Born in Middle Germany, he spent the first part of his career in the Dominican cloister in Erfurt as prior.

As a theologian, he taught at the Sorbonne in Paris two times a year, an honor granted only to him and to Thomas of Aquinas. As a preacher, he worked above all in convents in the Rhineland. Biographical or autobiographical accounts of his mystic experiences have not been handed down to us. However, they are accessible from his sermons that were written down by nuns.

Meister Eckhart, probably the most important German mystic of the Middle Ages, expressed the mystic experience of time in his German sermons in an impressive way: purified by an inner letting go and a "becoming empty," in a moment he steps out of the stream of flowing time and enters into God's eternity. It was an "eternal now" in which all time was settled. The human soul makes possible his transcendence of the time of eternity. As an "image" of God, it had been created as a mirror: in unlimited breadth it could pick up and reflect everything, even the light of God, when it has disposed of all other things. On the basis of our souls, "at the bottom of our souls," or in the "sparks of our souls," we can now reflect the eternal, and so we can experience the mystic moment.

The aim of all Christian life, according to Eckhart, was the representation of holy history in the "here and now" of life. For him it was about the journey from time into eternity, and it was about the return trip from eternity into time. This return trip qualified everyday time as "in God" (Kunz 1985).

The structure of his mystic experience corresponded with his three-fold gifts, mentioned above: journey to God, God's birth in the soul, and the return trip into the active life. In this sense Eckhart could preach (Quint 1963, 345-46):

> It is a strength in the soul, the reason, that from the outset, as soon as it becomes aware of God or tastes him, has five qualities in itself. For the first, it gives up the here and now. For the second, that it resembles nothing. For the third, that it is heard immaculately. For the fourth, that it is acting or searching in itself. For the fifth, that it is an image. Firstly, it replaces the here and now. "Here" and "now" mean time and place. "Now," which at the very least is in time, is neither a piece of time, nor a share of time; probably however it is a taste of time; it is a mountaintop of time and an end of time. Still, as small as it may be, it must depart; everything that touches time or has a taste of time must depart.

However much Eckhart emphasized temporality as a hindrance to the knowledge of God, it is very clear that for him this tendency did not grow into escapism. The biblical belief in creation was now sufficiently strong, after the journey to God — out of time — for the return trip to reenter the world.

Meister Eckhart was ahead of his time, and he has not always been rightly understood. Thus, Pope John XXII, who showed his resistance to various innovations through numerous prohibitions, suspiciously viewed twenty-eight of Eckhart's statements as heresy. Eckhart's tracks are lost in Avignon, whence he had hurried to this pope to defend in vain his theology from the accusations of the Inquisition. His end lies hidden in darkness.

Angelus Silesius (1624-77)

This witness also lived in a time of change. In the stormy times of the Counter-Reformation, he was converted to the Catholic faith, although he was originally a Protestant scholar and doctor. A step such as this aroused great attention at that time. Johann Scheffler, the "heavenly messenger of Silesia" (Angelus Silesius), in his "Cherubic Pilgrim," had condensed mystic experiences and wisdom into short couplets, as in small focal points. Thus, he said about time and eternity:

> I myself am eternity when I leave time
> And I am united with God and God with me.

> There is nothing that moves you, you move yourself up to the wheel, which moves of itself and has no rest.

> You yourself make time, which clockworks are intending,
> If only you restrain unrest, then is the time to come.

Here it is clear that the mystic experience of time aimed at overstepping time and everything temporal. The sphere of the divine in itself was "timeless," and it stood, as it were, perpendicular to the flow of time that sought in vain to merge with it. The condition of the possibility of experiencing God was "leaving" time, "stopping" the rolling wheel of time and the "restraint" of unrest that came from the sphere of the temporal.

Friedrich Schleiermacher (1768-1834)

Schleiermacher lived in the flowering of the German Enlightenment. In his lectures *On Religion* in 1799, Schleiermacher distanced himself from rigidly orthodox, rationalistic Protestant orthodoxy and Kantian metaphysics. In connection with the immortality of the soul in *Time and Eternity,* Schleiermacher gave his opinion, and he thought of a genuinely mystic understanding of time. He wrote near the end of the section entitled "Essence of Religion" (1994, 100): "If our feeling nowhere attaches itself to the individual, but if its content is our relation to God wherein all that is individual and fleeting disappears, there can be nothing fleeting in it, but all must be eternal. In the religious life then we may well say we have already offered up and disposed of all that is mortal, and that we actually are enjoying immortality."

In this context he criticized the wish of many Christians for individual immortality, in the sense of an extension of the individual soul beyond death — reminiscent of Meister Eckhart and medieval mystics — as he wrote (1994, 100-101):

They desire an immortality that is no immortality. They are not even capable of comprehending it, for who can endure the effort to conceive an endless temporal existence? Thereby they lose the immortality they could always have, and their mortal life in addition, by thoughts that distress and torture them in vain. Would they but attempt to surrender their lives from love to God! Would they but strive to annihilate their personality and to live in the One and in the All! Whosoever has learned to be more than himself, knows that he loses little when he loses himself.

Schleiermacher continued with the famous formulation, which was literally taken from the mystic experience of time within Christianity (1994, 101): "In the midst of finitude to be one with the Infinite and in every moment to be eternal is the immortality of religion." So far we have mentioned only records from the Christian heritage. But the mystical experience of timelessness is also well attested in other religions, such as *Ksana* in Hinduism, *Nikon* and *Uji* in Zen-Buddhism, and *Waqt* or *An* in Sufism (Achtner 2001, 387).

SUMMARY

Up to now we have outlined the possible human experiences of time: the cyclic-mythic, the rational-linear, and the holistic-mystic experiences. The first and largest part of our investigation is concluded.

We have seen that each of these three structures of time is connected in distinctive ways with exogenous time. The simplest configuration occurs when the cycles of weak mythic consciousness move in concert with the rhythms of nature. Or, to put it technically, the simplest case is when endogenous and exogenous time swing in synchronism with double-sided cycles.

We have seen further that the rational-linear experience of time has no counterpart in nature, since nature is dominated by periodically recurring processes. The rational-linear experience of time led to an asynchronism between endogenous and exogenous time. The greater stability of ego systems helped obtain a greater distance from nature — it became the object — and it enabled humans to take part in cultural affairs. A synchronism with exogenous time happened in a roundabout way, since the clock was a product of culture. This synchronism with exogenous, social time — the time of culture, so to speak — required humans to adapt in ways that were limited biologically. For example, rational-linear time could not be changed into endogenous-cyclic time (jet lag; shift work). Another example was that the possible effects of acceleration, which are connected with rational-linear time, could not be increased as you like.

Finally, it is not clear whether the mystic experience of unity deals with an actual experience of existence or whether it is merely an experience of oneself. If the first is the case, then we could speak of a synchronism between endogenous and exogenous time — on a higher level.

At this point we have reached the second pole of our tri-polar framework. First, we must isolate the methods from the total framework and observe and ask about the special place of time in nature. This point must be clear: every conception of time in nature relies on definite philosophical and religious ideas about nature. The Aristotelian concept of physics that is governed by entelechy is quite different from that of natural science that is geared to nominalism. However, the latter's concept of an objective natural world differs from the feelings of German romanticism about nature.

In the following chapter, we choose the perspective of our European scientific view since its concept of time has largely dominated our lives and has also conquered the globe. This choice naturally reduces our tri-polar system to its historically developed self-understanding *(etsi deus non daretur)*. Additionally, it seems to us a quite remarkable process. In the natural sciences, the time of nature is not promoted by nature itself — Newton had banished all natural processes from his definition of time. Rather, it was drawn from our rational-linear time already known by the description of external, idealized parameters. It is time without qualities (standardized and homogeneous), and it corresponds to a concept of matter without qualities. Matter is a manipulable, amorphous mass, and its qualities are attributed by science to quantities. The self-meaning and the times of nature itself were sacrificed to the concept of standardized time and standardized nature. Therefore, in our contemplation of the time of nature, we have to do with a "prepared time" and with a "prepared nature." In our considerations that follow we must always keep this double reduction in mind: the reduction of the tri-polar system to the pole of nature, and the reduction of the inner natural time to the external parameter of standardized linear time.

However, in the course of the story of natural science it will probably turn out that this idealized concept of time itself, in an astonishing way, also developed a wealth of structure closer to nature.

The Time of the World

THE SCIENTIFIC VIEW OF TIME

The time of the world, the time of nature that surrounds us, corresponds in the framework of the tri-polar scheme to exogenous time, and in the latter to natural time. Here, the concept of time that underlies the modern sciences — especially physics — will be introduced, and its immanent structure will be shown. What are the fundamental qualities of scientific time?

The following points are addressed:

- The symmetry of time in scientific laws;
- The injured symmetry in the second law of thermodynamics;
- Time in quantum mechanics;
- Time in relativistic physics and in cosmology; and
- The significance of time for chaotic systems.

NEWTON'S VIEW OF TIME

Sir Isaac Newton's (1642-1727) work was fundamental for the development of the modern sciences and also for the modern concept of time. (His chief work appeared in 1687, *Philosophiae Naturalis Principia Mathematica*, hereafter cited as Newton 1988 [German translation] and Newton 1952 [English translation]). In it he gave serious thought to the concept of time. This consideration mostly took place in the context of essays on absolute space since for Newton time was essentially connected with the concept of space, and it was dealt with in analogy to space.

Newton's Definitions and Distinctions in Time

At first, Newton divided the concept of time and distinguished between *absolute time* and *relative time*. These two will be arranged in the tri-polar scheme, to which they are first introduced. Newton dealt with them individually (1952, 8):

> Hitherto I have laid down the definition of such words as are less known, and explained the sense in which I would have them to be understood in the following discourse. I do not define time, space and motion, as being well known to all. Only I must observe, that the common people conceive those quantities under no other notions but from the relation they bear to sensible objects. And thence arise certain prejudices, for the removing of which it will be convenient to distinguish them into absolute and relative, true and apparent, mathematical and common.
>
> I. Absolute, true, and mathematical time, of itself, and from its own nature, flows equably without relation to anything external, and by any other name is called duration: relative, apparent, and common time, is some sensible and external (whether accurate or unequable) measure of duration by the means of motion, which is commonly used instead of true time; such as an hour, a day, a month, a year.

He carried out the connection between absolute and relative time (1952, 9):

> Absolute time, in astronomy, is distinguished from relative, by the equation or correction of the apparent time. . . . It may be, that there is no such thing as an equable motion, whereby time may be accurately measured. All motions may be accelerated and retarded, but the flowing of absolute time is not liable to any change.

Both concepts of time had to be strictly separated, and they should not be confused, for that is how errors arise, as he noted (1952, 12):

> Wherefore relative quantities are not the quantities themselves, whose names they bear, but those sensible measures of them (either accurate or inaccurate) which are commonly used instead of the measured quantities themselves. And if the meaning of words is to be determined by

their use, then by the names time, space, place, and motion, their [sensible] measures are properly to be understood; and the expression will be unusual, and purely mathematical, if the measured quantities themselves are meant. On this account, those violate the accuracy of language, which ought to be kept precise, who interpret these words for the measured quantities. Nor do those less defile the purity of mathematics and philosophical truths, who confound real quantities with their relations and sensible measures.

With this separation between relative and absolute "measurements," Newton created a presupposition about the actual objects of the new physics, namely, movements and the powers of causes. He sought to explain that in absolute space, in contrast to relative space, movement could only be changed by the action of a force.

Newton began with an abstraction from absolute quantities since the laws of mathematical and physical movement postulated by him did not correspond with observations. This correspondence was not possible for him since he did not know about the influence of friction on mechanical movement. The coincidence between the mathematically formulated "natural law" and experiment was achieved by transferring the observation from relative to absolute space. It is remarkable not only that Newton carried out this step formally, but also that he gave absolute quantities — very weighty — textual significance!

As evidence of authority for the existence of absolute time as well as absolute space, Newton could not refer to accessible data and observations. Therefore, in *De Gravitatione* Newton connected the existence of absolute measurements with God's existence (cf., for example, the corresponding implementations in Schneider 1988).[1] Essentially they were connected with Newton's image of the transcendental God. In *Scholium Generale* Newton elaborated on his image of God (1952, 370):

> He [God] is eternal and infinite, omnipotent and omniscient; that is, his duration reaches from eternity to eternity; his presence from infinity to infinity; he governs all things, and knows all things that are or can be done. He is not eternity and infinity, but eternal and infinite; he is not duration or space, but he endures and is present. He endures forever,

---

1. An attempt at experimental proof is presented in the "bucket experiment" (see Newton 1988, 49-50).

and is everywhere present; and, by existing, always and everywhere, he constitutes duration and space.

Further considerations of Newton's image of the transcendent God, especially in the context of dualistic ideas of space and time, will be found in Achtner 1991, 54ff.

Within the tri-polar structure of time, Newton's relative time corresponded to subjectively perceived time, which was endogenous time. Absolute time was an implementation of its origin in the existence of the transcendent God. However, this invariably running time that was distant also had significance for our world, for it formed the time scale on which all natural laws run without change. In other words, absolute time was a projection of transcendent time onto the idealized, measurable, and objective time of science, that is, exogenous time. With its two concepts of absolute time and relative time, Newton's structure of time was a reduction of the tri-polar structure of time.

Quantified Time

Surprisingly, Newton's idea of space and time was clearly formed from a quantum structure: not only matter but also absolute space and absolute time were quantified. Intellectually, he started from a quantization of matter, which existed in the smallest particles. These particles could only occupy discrete portions of space, and therefore space had to be quantified. Consistently, he then proceeded to the "smallest portions of time," quanta of time, for which he frequently argued in his *Principia*. For example: "As the order of the parts of time is immutable, so also is the order of the parts of space" (1952, 10).

Newton used these portions of time argumentatively and in the reasoning in many places in his *Principia*. For example, in connection with centrifugal force, he set forth the premise that the observed movement "precisely describes in the smallest portion of time an emerging curve" (1988, 97). Newton used these "smallest portions of time" in arguing for early considerations of calculus (see, e.g., Laugwitz 1986). It was a further central distinction between absolute and relative time, and it was an essential distinction from the continuum mechanics of Leibniz.

For a Supplement: Einstein's Concept of Time

Albert Einstein (1879-1955) met a very similar distinction when he separated *subjective* and *objective* time. Subjective time was an individual's "self-time," and as such it was not measurable (therefore corresponding to endogenous time). Objective time was centrally connected with a clock for measuring time (exogenous time of the sciences). His strongly operational definition of time was oriented by the measurability of time. This definition essentially took into account the finiteness of the speed of light, and it took into account the resulting problem of "simultaneity" that resulted from it. He put it this way (1922):

> For measurement of time, we think of a clock "U" arranged somewhere resting against K. But with help of this clock, events cannot be evaluated directly with regard to time if the spatial distance from the clock is not negligibly small. That is because there are not any "signals of the moment" at its disposal in order to compare these events timewise with the clock U. One can use the constant of the speed of light in a vacuum as a complete definition of time.

The central significance of the concept of the clock was emphasized again in Einstein's definition of time.

In our tri-polar system, we can clearly identify this concept of time with exogenous time.

THE SYMMETRY OF TIME IN THE LAWS OF PHYSICS

The laws of classical physics show the following symmetrical behavior:

- The classical laws of mechanics and electrodynamics are invariant with the reversal of time.

It is a surprising fact, considering their universality, that this principle continues in practically all areas of physics — with exception of the thermodynamics. It follows from the fact that, up to the time of quantum field theories, the natural laws yielded to a principle of variation. The starting point of our consideration is Hamilton's principle (cf., e.g., Goldstein 1985):

- The movement of a mechanical system is such that the linear integral

$$I = \int_{t_1}^{t_2} L dt$$

is a limit for the path followed. L = T−V, which is the difference between kinetic and potential energy.

Many known regularities of physics are equivalent formulations of this principle, as, for example, Fermat's principle (optics) and Maupertius's principle[2] (principle of the least effect).

The calculation of variations shows that Hamilton's principle is equivalent to the so-called *Euler-Lagrange equations* (3.1):

$$\frac{d}{dt}\frac{\partial L}{\partial \dot{q}} - \frac{\partial L}{\partial q} = 0$$

L designates = L $(q, \dot{q}, t)$ = T−V as above in the so-called Lagrange function, q the (generalized) coordinate and $\dot{q}$ the accompanying impulse (cf., e.g., Goldstein 1985; Fick 1988). With the substitution t →−t the following transformation takes place:

$$\frac{d}{dt} \to -\frac{d}{dt} \text{ and } \dot{q} \to -q \Rightarrow \frac{\partial}{\partial \dot{q}} \to -\frac{\partial}{\partial \dot{q}}$$

The *same* equations (3.1) arise! The laws gained from (3.1) are therefore invariant with a reversal of time![3]

In classical mechanics, the symmetry of time is directly obvious. Newton's equation of movement "power is change of impulse" is read formally as

$$F = \frac{d}{dt} p = m\, \ddot{x} = m\frac{d}{dt}\frac{d}{dt} x$$

The reversal of time t → −t now yields the same equation of move-

---

2. Maupertius in 1747 conceived of the principle of "least work" quite vaguely based on theology(!). Euler and Lagrange achieved an objective basis.

3. The same circumstances with a different observation result in an interesting picture: E. Noether shows generally that a symmetrical transformation implies a conserved quantity (constant movement).

ment if the well-known rule of the rake for the extraction of a range (here the −1) is used for the differential operator:

$$F = m \frac{d}{d(-t)} \frac{d}{d(-t)} x$$

$$= m \frac{1}{-1} \frac{d}{dt} \frac{1}{-1} \frac{d}{dt} x$$

$$= m \frac{d}{dt} \frac{d}{dt} x$$

It should be noted that equations of the form (3.1) are essential for all areas of physics (disregarding empirical thermodynamics). Thus, the equations of the quantum field theory take the form, for example, of an Euler-Lagrange equation

$$\partial_\mu \frac{\partial L}{\partial(\partial_\mu \Phi)} - \frac{\partial L}{\partial \Phi} = 0 \qquad (3.1)$$

which is formally equivalent to (3.1) (L = L($\partial_\mu \Phi$, Φ), which designates Lagrange's density of the fields). As to L = ½(($\partial^\mu \Phi$)($\partial_\mu \Phi$) −$m^2 \Phi^2$), the Klein-Gordon equation ($\Box$ + m^2)Φ = 0, for example, follows, which describes particles with spin 0 and was treated as an early forerunner of the relativistic Dirac equation.

An Example

As an example, we consider a particle with potential: V(q) = k · q^2.

With kinetic energy T($\dot{q}$) = ½m$\dot{q}^2$

the Lagrange function follows L($\dot{q}$, q) = ½m$\dot{q}^2$ −kq^2.

With the equation (3.1), it follows further:

116

$$\ddot{q} + \frac{k}{m} q = 0$$

$$\Leftrightarrow \ddot{q} + \omega^2 q = 0 \ \textit{with} \ \omega = \sqrt{\frac{k}{m}}$$

It is the equation of movement for a harmonic oscillator with the well-known solution

$$q(t) = A \cos \omega t + B \sin \omega t$$

But equally, movement in the reversal of time also fulfills the equation of movement!

$$\tilde{q}(t) = \tilde{A} \cos(-\omega t) + \tilde{B} \sin(-\omega t)$$

This situation corresponds to the first solution with the initial conditions

$$\tilde{A} = A \ \textit{and} \ \tilde{B} = -B$$

THE SECOND LAW OF THERMODYNAMICS

Hardly any other physical law is open to more diverse explanations than the second law. Statistical physics has a particular significance for explaining the second law. On the basis of atomistic consideration, Ludwig Boltzmann (1844-1906) succeeded with an expansion of the concept introduced by Claudius: entropy. Boltzmann designated his "statistical entropy" as *H-function* (cf. Boltzmann 1896). Zucker (1974) was somewhat close to it.

The main statement of thermodynamics gives expression to empirically gained realizations about thermodynamic behavior. The first law is the thermodynamic formulation of the conservation of energy:

$$dU = \partial Q + \partial A$$

To put it into words: a change of inner energy corresponds to the heat supplied to the system and to the work performed in the system.

The second law is explained, likewise, as follows:

- By itself, a system never changes to a significantly less probable condition.

We want to expound this situation on the basis of an example: four molecules of gas (numbered from 1 to 4) are in container A. Container B, which is the same size and initially empty, will be connected to container A in such a way that gas molecules can travel between the two containers. We now want to count the number of possibilities that (a) all molecules are in B or (b) two molecules of gas are in A and two are in B. For (a) there is exactly one possibility (A = empty, B = 1,2,3,4). For (b) however, there are 6 possibilities (A = 1,2 and B = 3,4; A = 1,3 and B = 2,4; A = 1,4 and B = 2,3; A = 2,3 and B = 1,4; A = 2,4 and B = 1,3; A = 3,4 and B = 1,2). The "more disorderly" condition appears far more frequently. A system left to fend for itself will probably never reach the condition of "all atoms in one container." This effect becomes even clearer if the number of particles is increased (4 molecules are a completely unrealistic view, when one has on hand the magnitude of a mole of molecules [a gram molecular mass], which is $6 \cdot 10^{23}$ molecules), since probability is a function of the capability of the number of particles.

We want to introduce even further formulations of the second law:

- There are irreversible processes.

This statement says that the processes in nature flow in a direction established by the second law. There is an appreciable distinction from the microscopic symmetry of time described in the prior section: point a video camera at (a) a (quasi-frictionless) stroke of two billiard balls and (b) a waterfall, then with (a) one cannot decide whether the tape runs forward or backwards, but with (b) everyone expects that the water falls downward and cannot fall upward.

- There is no perpetual motion of the second type

A perpetual motion of the second type is an appliance that does not violate the law of the conservation of energy, but it performs work in that it cools down a cold body further and warms up a warm one further. However, every irreversible process could then be made reversible with it.

Open and Closed Systems

These formulations of the second law are valid only in *closed* systems. If there are effects from outside this system, then an interaction happens with its surroundings and the system does not behave as described above. In the example above, if mechanical equipment periodically collects all the atoms of gas in a container, then the condition of uniform distribution is no longer the most likely. In order to change it into a more improbable (ordinary) condition, a system must be supplied energy from the outside. Therefore, the question is whether our world is an open or a closed system. That is, in the case of a closed world, it moves toward an ever "more disorderly" condition (the death of the universe by heat).

Entropy and the Second Law

Boltzmann introduced the entropy S as a measurement for the probability of a condition on the basis of statistical physics. With it, the second law can also be formulated this way:

- A system only changes in such a way that its change of entropy is larger than or is exactly zero.

There are different mathematical formulations for entropy. Kolmogorov's definition of entropy is the best clarification of the concept. In the simplest case, it is $S = k \cdot \ln p$. The probability for a condition is included in p, and k is the Boltzmann constant.[4] The logarithm leads to the additive structure of entropy: the entropy of a total system is the sum of the entropies of the subsystems since the total probability results from the multiplication of the separate probabilities. Then the laws of logarithms provides $\log p_1 \cdot p_2 = \log p_1 + \log p_2$.

This approach was continued by Jaynes (Jauch and Baron 1961) and Prigogine (1970), among others.

> *Entropy as the thermodynamic function of condition*
> In conclusion, the concept of the *thermodynamic function of condi-*

---

4. $k = 1.38062 \cdot 10$ J/K.

tion should still be dealt with: it is a function, which is unequivocally established by a few values of conditions (e.g., pressure and temperature). This function is independent of how these values were gained, and therefore it is independent of the prehistory. The supplied quantity of heat Q and the work performed A are not such functions; on the other hand the inner energy U and especially the entropy S have mathematically long-range consequences (especially for circular processes).

In differential notation, it is expressed by the notation dU for the total differential and δQ for the path dependent change (see the formulation of the first law). That entropy is a function of a condition is a central realization of thermodynamics. Vice versa, absolute temperature T can be defined (!) as an *integrating factor*, as in the expression:

$$dS = \frac{\partial Q}{T}$$

Directional Time

With the help of entropy, the second law can now be formulated mathematically:

$$t_2 > t_1 \Rightarrow S(t_2) \geq S(t_1)$$

However, the direction of time is firmly predetermined, and this exogenous time runs in a distinctive direction! Many times the concept of the arrow of time is used in order to express the directionality of the course of time (see, e.g., Burger 1986; the expression "the arrow of time" was introduced by Arthur Eddington in 1928 in "The Nature of the Physical World").

Directional Time and Causality

Directional time is also quite important for causality. The essential foundation of all modern sciences is that the cause must be before the effect, and that presupposes a direction of time. More about causality is found in Mittelstaedt 1963. Directional time and causality also provide a distinction between past and future, which Schiller aptly characterized with the words:

120

The pace of time is threefold
Hesitantly, the future draws near
Quickly as an arrow the now has flown
Eternally quiet, the past stands still.

In modern quantum mechanics, the postulate of absolute causality has been put in doubt within certain limits (see, e.g., "Die Quantenphysik der Zeitreise" ["The Quantum Physics of Time Travel"], in *Spektrum der Wissenschaft* [November 1994]).

Summary

We have seen that the structure of exogenous natural time, first of all, is infinite and linear, and the second law gives it a direction.

This basic scheme is practically applicable to the whole of physics. For the concept of time, we want to introduce two particular areas of physics in more detail: quantum mechanics and the theory of relativity.

TIME IN QUANTUM MECHANICS

The history of quantum mechanics begins with Max Planck's attempt to reconcile classical dynamics with the second law (cf., e.g., Prigogine 1970). With the development of the Bohr-Sommerfeld model of the atom, quantum mechanics was separated from thermodynamics. We want to discuss here only briefly, and insofar as a particular viewpoint can be gained for the concept of physical time, the development of quantum mechanics, its particular performances, and its gain in realization — somewhat concerning the measuring process. For another and more elaborate overview see, for example, Hund 1984. For the theory itself see Fick 1988.

Time is entitled to a special role in quantum mechanics. It differs conceptually from what is *observable*, the measurable quantities, because it is (only) a parameter of the understanding of quantum mechanics. That presents a special problem in connection with the second law and irreversibility since a pure parameter does not admit any distinctive direction of time. This matter is still in discussion today, and it is one point at which philosophy and physics meet.

The Schrödinger Picture

Schrödinger established a formulation of quantum mechanics, and it was derived from an equation of a wave from an analogy to classical optics. The essential quantity is thereby the vector of condition ψ (x, t), a complex variable, whose absolute square in $|\psi|^2 = \psi * \psi$ describes a probability.

The Schrödinger picture describes dynamics by the Schrödinger equation:

$$i\hbar\frac{\partial\Psi}{\partial t} = H\Psi \tag{3.2}$$

On this occasion, H represents the Hamilton operator. We assume here that the Hamilton operator is not explicitly dependent on time.

The operator is unchanged in this picture during the time in which the condition ψ (x, t) changes with regard to time. The dynamics therefore is reproduced by the change of the function of condition.

Determinism and Symmetry in the Schrödinger Equation

If the condition of function ψ is known at a certain time, for example t = 0, then from (3.2) the conditions for any prior and later times can be calculated; in this way a complete determinism prevails. However, the situation of classical physics differs fundamentally from quantum mechanics since in the latter the function of condition ψ is only a statement of probability!

Equation (3.2) is a partial differential equation of the first order in time. At the time t = 0, if you have the known solution ψ (x, 0), it becomes formally

$$\Psi(x,t) = \exp(-iHt)\,\Psi(x,0) \tag{3.3}$$

which is solved for any time t. With the reversion of time, t → −t, which is simply the consequence that in (3.3) t is replaced by −t and therefore past and future are exchanged!

The Heisenberg Picture

In the Heisenberg picture, the basis of the Schrödinger picture is continued, and one proceeds from a constant function of condition with regard to time, but with variable operators with regard to time. For operator A, the equation of movement in the Heisenberg is valid:

$$i\hbar\frac{\partial A}{\partial t} = AH - HA = -LA \qquad (3.4)$$

Also this equation may be solved formally, in analogy to (3.3). The observable variable time A (t) is of the form

$$A(t) = \exp(iHt)\, A \exp(-iHt) \qquad (3.5)$$

For a transformation of time t → −t, it has the same value as in the Schrödinger picture, as past and future trade their meaning.

Further Pictures

Besides the Schrödinger and Heisenberg pictures introduced earlier, there are still "mixed solutions" that are derived from changed conditions ψ, as well as from observable and changed A. Thus, the two pictures presented here represent extreme approaches. However, what do these approaches mean for "real physics," for measurements, for experiments?

Application of the Pictures

Already the concept *observable* has occurred without being explained in detail. In terms of quantum mechanics an observable is a measurable quantity. Mathematically they are described as "Hermitian operators," which have as a consequence that their "measurements" provide real — and not complex — measurements. A system is found in the condition ψ, and we want to measure the quantity A. That means we want to determine the expected value "A." In terms of quantum mechanics, the integral

$$\int \Psi^* A\Psi d^3 x$$

must be evaluated. The equivalence to the observed pictures is clear since the evaluation of the integral expression leads to the same result — therefore to the *same* measurement!

TIME IN THE THEORY OF RELATIVITY AND IN COSMOLOGY

In as complete as possible a representation of the main discussion of the first two sections of this chapter concerning the concept of physical time, we want to add a summary of the meaning of time in relativistic physics and in the development of the cosmos.

Time in the Special Theory of Relativity

In 1905, when Albert Einstein introduced the special theory of relativity (STR), he essentially broadened the understanding of physics. We are reminded of the well-known remark of Minkowski, when an attendee of his lecture asked him to think no longer of space and time but rather to look at four-dimensional space-time.[5] What is meant by that?

As an essential postulate, the theory of relativity assumes that the (finite) speed of light c (approximately 300,000,000 meters per second) is the greatest speed that an informational signal can attain. The question has been asked as to which place x' is observed as a part of a certain time t' from a frame of reference that moves with the constant speed v.[6] In a frame of reference at rest in the place x and for the time t, the problem was once classically solved by the Galileo transformation. It is shown in the (one-dimensional) form in figure 6 (p. 125).

A relativistic athlete runs with the speed of 0.75 of the speed of light toward a railroad train that is moving at 0.75 of the speed of light. The rate of closure of the athlete with the train would be 1.5 times the speed of light. That would be half again as fast as light, which is why this postulate is violated. Therefore, the Galileo transformation is false for large speeds!

The heart of the STR is the so-called Lorentz transformation, which is the correct relativistic form of the Galileo transformation:

---

5. The following presentation does not follow the historical route: Einstein's work in 1905 observed electromagnetic phenomena. The great influence of the vain attempts concerning ether and the Michelson-Morley experiments is essential and preparatory for further meaning.

6. Two such systems, which move with constant velocity, can be called inertial systems.

Figure 6: Transformation between intertial systems

$$x' = x - v \cdot t$$
$$t' = t$$

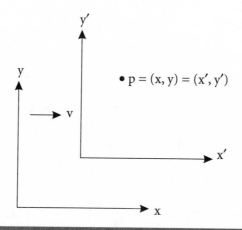

$$x' = y(v)(x - vt)$$
$$t' = y(v)(t - vx/c^2), \qquad\qquad (3.6)$$

$$whereby \quad y(v) = \frac{1}{\sqrt{1 - v^2/c^2}}$$

The second line of (3.6) is part of the central peculiarities, in that other time scales prevail in the different frames of reference. This is the appreciable distinction from the Galileo transformation, which proceeds with the same time scales in the system at rest and in the system in motion. This leads, for example, to the phenomenon known as the paradox of twins (twins meet, one is young on the basis of travel at "relativistic speeds," and the other is old). In a borderline case, as the speed of light approaches infinity, mathematically the Lorentz transformation turns into the Galileo transformation.

In this way, space and time are connected in a four-dimensional spacetime, which is what Minkowski meant with his remark.

There are numerous consequences of this connection. For example, the concept of simultaneity in this four-dimensional structure is complex: two events that are simultaneous for an observer in one particular inertial system could be not simultaneous for an observer in a second inertial system. This situation is not possible in a physics that is not relativistic.

The Concept of Horizon

Another consequence of the STR should be discussed. According to a physics that is not relativistic, a particle that at the time t_1 is at the place x_1, for the time $t_2 \neq t_1$ is at any place x_2. It is not so in relativistic physics: Since the speed of light is an upper limit of all expansion, then $|x_2 - x_1| \leq c$ times $|t_2 - t_1|$ must be valid. The event horizon of the particles at (x_1, t_1) are points of space-time ("beams of light") $(x_1 + ct, t_1 + t)$ — whatever lies beyond these points has no interaction with the particle!

A mathematically well-rounded and modern presentation of the STR can be found in Sexl and Urbantke 1987 or Rindler 1977.

The General Theory of Relativity

In contrast to the STR, the general theory of relativity (GTR) takes into account the effects of gravity (Einstein #1922). There is also an influence on the scale of time: In a strong field of gravity time appears to pass more slowly than in a weak field. The theory in itself is essentially described by Einstein's equations of the gravity field. Their solution works only for particular borderline cases: for example, the Schwarzschild solution, which gains its central significance by the Schwarzschild radius, and black holes. For the GTR see, for example, Rindler 1977; for the meaning of time see Mittelstaedt 1989, chapter 8.

In contrast to the STR, the GTR is regarded as not closed. To be sure, the (nonclassical) effects of the GTR have been confirmed magnificently in many experiments. However, up to now a quantization of the GTR has not been accepted, and in its present form it is divorced from "the rest of physics." Also, the long fruitless search for waves of gravity predicted by the GTR is cautiously affirmed. Many newer beginnings (e.g., the theory of super strings) are under discussion.

Cosmological Consequences

Of course, the question of the future "of the universe" is obvious in this context. This discussion was essentially kindled by the discovery of Edwin P. Hubble (1923) that the light of a distant source is shifted to a lower frequency (red-shift: this is the relativistic Doppler effect). Hubble observed the red-shift in light from distant galaxies. In fact, the amount of this shift, and therefore the speed, is greater the farther off the galaxies are. It comes to about 17 km/s additional speed per one million light-years distance. This discovery led to the generally accepted assumption of an expanding universe (often explained by the expansion of a balloon). The singular origin of all expansion is the big bang, which happened some ten to twenty billion years ago.

The future of this expanding universe seems to make possible three scenarios:

- A universe that is always expanding: the *open universe;*
- A universe that is slowing in its expansion, and which will subsequently collapse: the *closed universe;*
- Albert Einstein and Willem de Sitter in 1932 proposed a third model, which lies between the other two: an *ongoing universe,* which is continually slowing its expansion and seemingly "coming to a standstill." (In this context, the concepts "open" and "closed" have a different meaning from that of the section "Open and Closed Systems.")

Which of these three scenarios our universe experiences is always a central component of current research. The answer to this question essentially depends on the density of matter in the universe: Is there enough gravity to bring expansion to a standstill? The critical value of the density of matter is approximately the mass of 3 protons per 1,000 liters of volume of space. The matter that emits light does not suffice to prevent the first scenario, and it amounts to only about 5 percent of the critical value. The density of the dark matter (e.g., black holes), which can only be appraised with difficulty, seems to be too small as well. The open question is the mass of neutrinos at rest. The neutrino is a kind of particle that is found quite frequently. However, it is not yet possible to determine its mass at rest. We only know that it is extremely small. Therefore, one can only say for the moment that the density of the universe is at least a twentieth of the critical density, and that at most it amounts to ten times the critical density.

The scenario of the big bang is of central importance for current physics today. The question as to what happened *before* the big bang is frequently discussed, but naturally such questions are predominantly of a speculative type. The model of the big bang brings up new problems. The obvious "shallowness" of space is not explained, and neither is its extraordinary homogeneity. Since around 1979, a refinement that has been discussed is the *inflationary universe* (at first under the name "Starobinsky model," which was developed at the Landau Institute for Theoretical Physics, Moscow). It broadened the original big bang model with a very short phase of extreme expansion, which does not correspond to current natural laws. This model explained, among other things, the minimal deviations from the homogeneity of 3-K background radiation. Through the new Cosmic Background Explorer (COBE) satellite experiments, this model was strengthened.

DETERMINISTIC CHAOS

In current scientific discussion, the concept of deterministic chaos is of special significance, and it leads occasionally to confusion about its meaning. In the following, the concept will first be explained, and then its meaning will be illustrated.

The literature in this area is quite diverse. Devaney's work (1989) is recommended as a good and extensive introduction that does not use too many mathematical tools. Wiggins (1990) takes the overview further, and Guckenheimer and Holmes (1986) carries it basically the furthest, but also critically.

The Definition

Chaos is a quite new concept in the current discussion. The topic was discussed for the first time through the diverse and basic writings of Henri Poincaré (1854-1912) at the turn of the century. Since these writings were so complex, they have only been pursued in about the last twenty-five years. The lack of the general availability of powerful computers was surely responsible for this delay. These computers made possible numerical "experiments" that led to the evocative work of Feigenbaum, among others.

Since the concept is still relatively new, there is not a firmly established definition. However, it is generally acknowledged, as stated in Devaney (1989); it is widespread; and it will be introduced here. Since frequently in this area an abbreviation or simplification does not provide the correct concept, the complete definition follows in its mathematical form. Afterward, we will expound on the content of these definitions.

Definition 3.6.1 (chaotic function):

The function $\Phi \, \Lambda \to \Lambda$. (Λ, Φ) is designated exactly as *chaotic* (whereby Φ is assumed to be constant and Λ is assumed to be a nonfinite and closed quantity), if

1. Φ has a sensitive dependence on initial conditions of Λ ;
2. Φ is topologically transitive to Λ;
3. the periodic orbits are dense in Λ.

On this occasion Λ is a metric space (metrics d) and Φ is constant.

Although it appears to be confusing at first, this definition will be described briefly in content.

The third point, "the periodic orbits are dense in Λ," said nothing other than the surprising fact that mostly order prevails in chaos: the orderly paths (periodic orbits) fill the tree such that in each small vicinity of any point there is a point of a periodic track (this means *densely*). The second point of the definition, "topologically transitive," means that in the course of its movement it covers all of the space. It follows that one cannot dismantle chaotic movement into "partial movements." These partial movements would be easy to understand because they would occur only in separate parts of space.

For the meaning of "chaos," the first condition is most important: "sensitive dependence on the initial conditions." To call in the mathematical apparatus, this means

$$\exists E \in IR^+ \, \forall \varepsilon \in IR^+ \, \exists y \in \Lambda \, \exists n \in IN : d(x, y) < \varepsilon \Rightarrow d(\Phi^n(x) - \Phi^n(y)) > E$$

for all $x \in \Lambda$. This means that for every point x there is from Λ a positive number E, for which the following is valid: in each small vicinity around x (with "diameter" ε) there is a point y, whose path in the course of movement is more distant than E from the path of x. This is the kernel of the concept "chaos," in this sense of the term! Everything scatters, and in principle a forecast is not possible, even if the beginning values are known very exactly (making ε as small as you will).

"Chaos" therefore is only conditionally chaotic, and it is especially subject to very strict mathematical laws. This situation is described by the adjective deterministic. A problem frequently occurs in the difficult verifiability of the abstract mathematical definition — especially since experimental science can work with it only quite conditionally. There are other, widely equivalent, numerical analyses that establish more practical foundations. Usually established with the help of Lyapunov's exponents, these analyses convert especially the first point of the definition into a measurement of the scattering of the curves of the paths.

Example: "Bernoulli shift" σ:

All this should be clarified with a characteristic example: the starting point of this example of one-sided Bernoulli shift σ is the quantity of all results that take the values 0 or 1. A result (a_n) is a sorted sequence of numbers, such as, for example

$$a = 0, 0, 1, 0, 1, 1 \ldots$$

The shift picture σ now cancels the first member of the sequence and pushes all the following numbers to a position "on the left":

$$\sigma(a) = 0, 1, 0, 1, 1 \ldots$$
$$\sigma(\sigma(a)) = 1, 0, 1, 1 \ldots$$
$$\sigma(\sigma(\sigma(a))) = 0, 1, 1 \ldots$$

After three more iterations, total ignorance prevails about the result! After a few steps in this simple illustration, there is a completely unpredictable result, which is typical of illustrations of chaos. The following are valid for this illustration:

The statement in 3.6.1 (Λ, σ) is chaotic.

The formal test of the individual points of the chaos definition will not take place here, as it can be found in Devaney (1989). The content of sentence 3.6.1 should still be explained in order to clarify the concept of chaos.

The illustration σ, presented above, can simply be conceived as a binary combination in which the leading digit is removed and the remaining digits are advanced. This picture follows a strict law, and it is deterministic.

130

If there is a sequence whose first m limbs are known, there is *complete* ignorance about the mth iteration: a prognosis of the behavior of the system outside the (m − 1)th iteration is impossible. The unpredictability of this behavior is called chaotic. Altogether, this picture is a "classic example" of deterministic chaos.

Current supplement:
The attempt here to define chaos (frequently designated as "chaos in the sense of Devaney") gets a surprising expansion with the work of Banks et al. (1992) (theorem 3.6.1):

If $\Phi: \Lambda \to \Lambda$ *is topologically transitive and the periodic orbits are dense, then* Φ *has sensitive dependence on the initial conditions.*

Consequently, the essential demand of sensitive dependence on the initial values is only one result of the remaining presuppositions! (For a closer look see Banks et al. 1992.)

Chaos and Time: Poincaré's Picture I

A system of continuous time is observed periodically at the times $t_n = t_0 + n \cdot T$. From the condition at the time t_n the following can be determined:

$$x(t_{n+1}) = f(x(t_n))$$

Now, two questions are frequently asked:

- What can be said about the long-term behavior of the system?
- What does it mean to say that the illustration is f chaotic?

The first question leads to the concepts ω (limit of quantity) and attractor. The second has the following consequence: although the system is strictly deterministic, the future can be predicted for a relatively short period. Since the initial conditions are never fully and precisely known (fuzziness), after a few iterations total uncertainty exists about the behavior of the system! (This became clear with the example of the Bernoulli shift illustration: In the beginning of a sequence of n members, after n + 1 iterations there is total uncertainty.)

Chaos and Time: Poincaré's Picture II

In conclusion, what is necessary for the meaning of a discrete picture should be discussed since all known natural phenomena are described by continuous (differential) equations and not by discrete (difference) equations. The interest in the latter can be explained with Poincaré's picture. Poincaré introduced this concept for the first time in 1899 in the framework of his examination of the problem of three bodies. We assume that the differential equation

$$\dot{x} = f(x), x \in |R^n, f \in C^r$$

has a periodic solution of x_p (t), and therefore a solution with the characteristic:

$$\exists T \ \forall t : x_p(t) = x_p(t + T)$$

We now observe a surface of $\sigma \in | R^n$ (generally: a manifold dimension [n−1]). It is chosen in such a way that x_p cuts the surface transversely. Then, $V \subseteq \Sigma$ exists so that the trajectories (starting in V) meet again at a time of the magnitude T. The (first) meeting point is Poincaré's picture P:

$$P : V \to \sigma$$
$$x \mapsto \Phi(\tau(x), x)$$

In this way we have found the sought-after discrete picture, which is described clearly by the behavior of the continuous system! In Poincaré's picture the concepts introduced can be used directly, and in this sense a continuous system can be chaotic.

In conclusion, the advantages of Poincaré's picture should be mentioned again:

- *Dimensional reduction:* now we only observe points of the dimension (n−1) instead of the points of R of the closing problem.
- *Global dynamics:* global dynamics can frequently be understood by numerical analysis of Poincaré's picture, and also the behavior of time (see, e.g., Guckenheimer and Holmes 1986).
- *Conceptual clarity:* many concepts, which are ponderous in the han-

dling of the usual differential equations, are simply transferable to the accompanying picture from Poincaré.

- *Mathematical advantages:* difference equations are mathematically simpler to deal with than differential equations. Furthermore, they were examined more thoroughly until now (see, e.g., Devaney 1989).

In order to find out more details on Poincaré's picture, see Wiggins 1990.

Chaos and Time: Delayed Bifurcations

(Local) bifurcations are central for the central meaning of the oft-quoted way of chaos. A starting point is a dynamic system described by a differential equation, which depends on an external parameter μ:

$$\frac{d}{dt} x = f(x, \mu) \tag{3.7}$$

The solution to certain initial values x (μ, x_0) is at first examined at fixed points, that is, with invariable conditions. These values are given for f = 0 because of (3.7), whereby the fixed points are easily calculated. Further analyses show the stable behavior of these fixed points. A bifurcation is a qualitative alteration of the behavior of the fixed point. For example, there is the saddle node bifurcation, which is described by the differential equation $x = M - x^2$. For negative values of the parameter μ there are no fixed points. For positive μ however there are two fixed points: $x_{F,1} = +\sqrt{\mu}$ is stable, and $x_{F,2}$ (μ) $= -\sqrt{\mu}$ is unstable; $\mu_c = 0$ is a point of bifurcation. The best-known scenario of bifurcation that leads into chaos is Feigenbaum's scenario of the logistical equation.

The infinitely slow transition from a bifurcation point is examined by means of nonstandard analysis (for the nonstandard analysis see, e.g., Laugwitz 1986), which can portray this ideal and slow transition mathematically. Both theoretically (Walter 1994) and experimentally (magnetization of a ferromagnetic probe — yttrium, iron, garnet — in a strong microwave field [cf. Walter, Rödelsperger, and Benner 1996]) bifurcations were entered and then delayed. There is a retardation of the bifurcation and then the unexpected behavior of a jump when an unstable fixed point is suddenly left. This behavior of a jump is not an indication of chaotic behavior, but it is only a forerunner on the way to chaos.

More Order or More Chaos? The KAM Theorem

After we briefly showed the significance of the modern concept of chaos, the question remains open as to whether the chaotic or the ordered condition of nature is more widespread. On the one hand, our daily observations show a high degree of order in the course of nature, but then we need only to mention the movements of the planets as disorderly. On the other hand, numerous processes are chaotic in their natural condition, for example, healthy brain currents.

Arnold, and later Moser, have continued to work on a theory of Kolmogorov that finally led to the important KAM theorem (see Mahnke, Schmelzer, and Röpke 1992; Lichtenberg and Lieberman 1983; Guckenheimer and Holmes 1986). Full of great meaning, this theorem examines the large class of nonintegrable systems.[7] First of all, the undisturbed, integrable system is examined. The KAM theory takes the phase of space into account. In many beginnings, nonlinear dynamics are generally described by the effect J and the angle α, which can be understood as the transformation of the impulses p and the places q in polar coordinates. The variable action of integrable systems with two degrees of freedom in space covers the surface of a toroid in the course of movement (cf. Lichtenberg and Lieberman 1983). However, one must remember that two different frequencies describe the rotation of the toroid itself and the dynamics of the rotation of the toroid. Periodic movements originate if the relationship of these two frequencies is a rational number (the accompanying toroid is called *resonant*), and in other cases quasi-periodic movements arise. If the measurement of the rational number is the real number 0, periodic movements would be an unlikely exception. Starting with the influence of small disturbances on these systems, KAM theory describes the opening of periodic trajectories and the formation of "accidental," chaotic curves of paths. Meanwhile, the quasi-periodic movements are hardly changed (and not qualitatively). Consequently, a "mixture" of ordered and chaotic paths is created. On this occasion, the curves of paths can have both kinds of positive measurement, and — to say it descriptively — they appear in large numbers.

7. Poincaré discovered that not every dynamic system has to be integrable. An integrable system is known as such when the accompanying Hamilton function is only dependent on the action of J. As a consequence, if this action is unchangeable with regard to time, then it is true that $dJ/dt = 0$: J is a constant of movement as in the example with energy.

The important statement of the KAM theorem consists of an elaborate description of the complex phases of space — this description normally occurs by means of Poincaré's picture. At first, it has to do with accidental movements that are exceptions (beginning with the improbable periodic movements). Then it deals with an increase in the disturbance that gains significance until the system finally shows chaotic behavior (details can be found in Prigogine 1995; Prigogine and Stengers 1993; in mathematical formulations in Mahnke, Schmelzer, and Röpke 1992; Guckenheimer and Holmes 1986; Lichtenberg and Lieberman 1983).

For systems with strong disturbances (these are frequently called "large Poincaré systems"), there are different approaches in the scientific discussion. A comprehensive solution is only accessible with difficulty on the basis of extraordinary mathematical calculations.

Outlook

In most recent times, chaos theory has increasingly been applied to the organization of dynamic systems in order to understand temporal development better. It has examined how the variation of certain control parameters affected the evolution of a system. Especially in medicine, the search has proceeded to find attractors that are available for cancer cell growth, brain waves, cardiac rhythm, and epidemiology. Complex feedback systems, as well as a corresponding fine-tuning of parameters, are present in glucose reduction in the immune system, in the control of blood sugar levels, and in the blood-clotting system.

In chaos theory, the concept of time opposes the linear concept of time of classical physics in only one insignificant extension. In so far as one can clearly distinguish between determinism and predictability, in the application of this concept of time to systems, one can speak absolutely of a conceptual extension. Dynamic systems are seen to have their "own time" for each system, or a "system time." That is, time exists in a relatively stable state, into which a system with a disturbance falls back again. The question now is, How great a tolerance can a system have and still find its way back into a stable situation? Expressed differently, each system can be assigned a certain plasticity of time. If a disturbance remains within the time of plasticity, then a system is again regulated in its own time. If, however, the disturbance is outside this width of tolerance, it cannot regain its own stable time. Either the system disintegrates pathologically, or it has the chance to develop; that is, the system evolves to learn to adjust to the new realities of its "characteristic time." This question of the quantitative relationship of its own time and an advance of evolution seems to us to be of great interest. It certainly is an area of research that is significant and pregnant for the future: that is, to explore systematically these characteristic times of systems. An example of this exploration could be an examination of the numerous feedback systems in the human organism. This point of view could provide support for the definition of health: health is a balance between order and chaos. Too much order leads to sclerotic solidification; too little order leads to disintegration.

SUMMARY

First of all, it was the goal of the section "The Time of the World" to make it clear that the idealized concept of linear time predominates in classical physics. In many areas (mechanics, electromagnetism, etc.), this concept helped in leading to unprecedented success. However, we wanted to show above how this idealized concept of time was broken open by the pressure of empirical research, as well as by the innovative power of theoretical and conceptual thinking from within, so to speak. A richer structure and a more complex concept of time found a place. We can identify following stages:

- In macroscopic systems, thermodynamics clarifies that time has a distinctive, irreversible direction (arrow of time). The direction of time can be read from the increase of entropy in such systems.
- The findings of quantum mechanics led to the functions of waves or conditions as statements of probability. The strict determinism of the linear concept of time either imposes ontological or theoretical perception onto interpretive qualifications.
- In the special theory of relativity, the concept of time is discovered in a threefold sense of negating its idealization. First, the idea of an absolute and universal time is discontinued, and in its place appears the multiplicity of the relative characteristic times of objects. Second, time is connected with the inviolable limit of the speed of light as a constant: by the spread of light there is a predetermined horizon that is in-

violable. Third, time is connected intrinsically with space in the four-dimensional continuum. These three aspects can be interpreted as negating the idealization, and thus the renaturalization, of the linear concept of time.

- In the general theory of relativity the connection of time with matter is given: subjectively, time in the vicinity of matter passes more slowly. Here is manifested a connection of time with matter.
- Finally, it became clear in chaos theory that one must strictly distinguish between determinism and predictability in a mathematical model for the description of feedback systems.

As a result, we established that the linear concept of time experienced a negating of idealization and an opening in the history of physics. Briefly, it has essentially gained structure.

This result opens up an interesting perspective on transcendent time for us. We may suppose that there is still a greater wealth of structure than we saw developing in the history of the linear concept of time in physics. It appears extremely important to us that this wealth of structure of transcendent time wants to inform the time of the world and also the time of humans. If we look at humans in the sense of chaos theory as self-organizing systems (religiously: humans as acting beings who follow the will of God; biologically: humans as beings with orientations to space and time in experiential and referential action), then in the temporal sense we could assign them different "characteristic times." These times have already been discussed as our three endogenous structures of time: the mythic-cyclic, the rational-linear, and the mystic-holistic. Now, in the next chapter, we will see that disturbances are imposed on the inner time system of humans by transcendent time. That situation forces them to develop endogenous time, and to develop socially. The Old Testament and the New Testament narrated the history of these divine disturbances.

The Time of God

THE THEOLOGICAL VIEW OF TIME

OLD TESTAMENT

Introduction

It is not the aim of this section to speculate about the existence of the time of God. That would presuppose knowing something of God's inner nature. Admittedly, theologians have again and again tried to fathom the inner life of God: for example, in the form of the immanent doctrine of the Trinity. The latter seeks to clarify the relationship of the three to one another. Also, there have been attempts to discuss God's relation to time on the basis of theological considerations: for example, in the theory of predestination.

Rather, the aim of this section is to discuss God's effects on the human experience of time. If we speak of God's time, then we are not speaking of it in the sense of its belonging to God's being, concerning which we do not want to speculate, but rather of God's effects from outside of us. These effects are recognized on the basis of the witness of the Bible. This approach somewhat corresponds to the theological position: *Deum cognoscere est beneficia Dei cognoscere* ("to know God is to know God's benefits").

First of all, when we speak theologically of the "time of God," we face an exegetical problem. In our heuristic principle, we need to find out which poles and textual qualifications predominate in the Bible with reference to our tri-polar system. After an examination of the OT and NT, we will turn to further theological considerations.

Exodus: Holy History

Primarily, we speak here of God's effects on people, and, according to the biblical conception, we presuppose that God created the world with regard to time: As God's creation, time has a definitive beginning and a definitive end. As God's creation, the world is not eternal. Nevertheless, the world is the place of God's acting, and the place where God demonstrates reliability and responsibility for humans.

In our discussion of the nomadic life of the patriarchs of Israel before the settlement in Canaan, we took note, conditionally, of the asynchronism between endogenous and exogenous time, and also the increasing dominance of the transcendent pole. We noted the thrust of the phenomenon of the openness of time, which was surprising and singular in comparison with the other Middle Eastern cultures. This thrust resulted in the substitution of the time-related, nomadic God for the space-related God of epiphanies.

Once again this tendency was increasingly met in the context of the discussion about the historicity of the Israelite religion. The crucial entry of Israelites into the light of history dated from the calling of Moses, the Exodus from Egypt, the wilderness wandering, and, as the climax, Moses' meeting with God on Mount Sinai and the reception of the Ten Commandments (Hebrew: the "ten words"). The text that is crucial for our question of Moses' meeting with Yahweh was handed down in Exodus 3. Now, biblical studies have made it clear that it is not an authentic story of events that occurred, but rather that there was a complicated process of the working of different sources, E and J, into one another (Schmidt 1988). The text of the story of the calling is found in Exodus 3:6, 14 [NRSV]:

6 He said further, "I am the God of your father, the God of Abraham, the God of Isaac, and the God of Jacob." And Moses hid his face, for he was afraid to look at God. . . .

14 God said to Moses, "I WILL BE WHAT I WILL BE." He said further, "Thus you shall say to the Israelites, 'I WILL BE has sent me to you.'" [NRSV, using alternative translation in footnote e]

In verse 6, the text presented a continuity between the encounter of God with Moses and the God of the patriarchs. However, then verse 14 decisively led out of pre-Israelite religion by way of God's self-predication. For our question concerning our tri-polar model, it is crucial that the story unequivocally placed the emphasis on transcendent time. It occurred as well in the totality of the story. It told of the shock of Moses at meeting God in verse 6, and also in verse 8 with the expression that God "came down," that is, God left God's transcendence. In addition it is remarkable that even the naming of Yahweh in verse 14 contained a temporal aspect. Again, it concerns the phenomenon of the openness of time that we have already met in passing with the patriarchs. Yahweh was open to the future time of people, and Yahweh wanted to participate in some way in human time. Primarily, God's acts are in human time and history. This view was expressed in the selection of the auxiliary verb *haya* in the name I WILL BE WHAT I WILL BE. The aspect of action in time was especially shown in the verb form that referred to the name of God. We can therefore say that the specifics of this Israelite faith in God were to be seen with God's real transcendence in reference to time and the future, that is, contingent actions in time. This textual qualification, together with the dominance of the transcendent pole, guaranteed the dynamic openness of time. It also guaranteed the permanent instability of this tri-polar system in the Israelite religion. Yet already the Greek translation of the Old Testament, the Septuagint (LXX), no longer understood this peculiarity of the Israelite religion. It translated the self-predication of God in verse 14 *'ehyeh 'aser 'ehyeh* (I will be what I will be) with ο ων, that is, the being ("I am who I am"). In this way the religiously motivated dynamism oriented to Yahweh's openness of time became the philosophically motivated static system of time. In the final analysis, the concept of being of Greek metaphysics defused and negated time and took on its own particular dynamics. Again and again Israel succumbed to the danger of the deactivation of the concept of God — thereby stabilizing the tri-polar system at the expense of the *openness of time.* We will meet this danger now in the context of the prophetic understanding of time.

The Prophetic Experience of Time

The Preexilic Prophetic Experience of Time in Canaan and the Deuteronomistic Preachers

The entrance of the Israelites into the culture of Canaan represented another important turning point in Israel's history because there was an encounter with mythic deities of nature. These deities were worshipped in Canaan, just as they were in Egypt and Mesopotamia. In the alternating cycle of agriculture, by means of our tri-polar system we saw a stable condition in Egypt's religion. It happened, first, because the asynchronism between exogenous and endogenous time was missing, and, second, because transcendent time was missing. Stabilization on the mythic level, however, presented a danger for the dynamics of our tri-polar system. How did the Israelites behave in the face of this danger of leveling?

In fact, the Israelites absorbed some of the mythology of Canaan in that region of Palestine where the Philistines, the Phoenicians, and the people of Tyre and Sidon lived. It was enriched with their ideas of creation. More importantly, however, they entered into the mythic life of the cyclical changes in agriculture represented in Canaan's religion, which led to a mixture of Israelite faith in Yahweh with the Canaanite worship of a female deity, Ashtaroth, a goddess of fertility (Judg. 2:13; 10:6; 1 Sam. 7:3-4), and with the worship of Baal (Judg. 6:25; Hos. 2:19; 13:1). However, this mixing of worship did not show up in a taking up of seasonal celebrations — as, for example, the harvest and wine celebrations — and the practices of sacrifice on the high places *(bamot)* in worshipping Asherah (Wolff 1969, 303ff.).

We could take the view that in this process of takeover and adaptation the religion of the Israelite conquerors could have been completely assimilated by the Canaanites — just as it happened again and again with many other processes of assimilation with mixtures of culture and religion. However, the takeover of these celebrations of the seasonal cycle by Israel was already connected with a new interpretation that was crucial. The natural-cyclic celebrations were historicized (Rad 1980), in that the Israelites undergirded them with the events of their own story, and they were consequently conferred with Israelite historical memory. Thus, for example, the feast of the harvest (Exod. 23:16) was undergirded with the Exodus from Egypt. The celebration of the wine (Lev. 23:42-43) was undergirded with the time in the wilderness. Nevertheless, this historicizing reinterpretation did not have the power to turn off the mythic "center of gravity" permanently. In the preexilic time, four prophets appear: Micah, Amos, Hosea, and, still later, Isaiah. By a massive criticism of cult and sacrifice, they strove to prevent the process of assimilation by upholding the horizon of history (Amos 5:21-27; Hos. 9:1-9; Isa. 1:10-17). We read in Amos 5:21-27 [NRSV]:

21 I hate, I despise your festivals, and I take no delight in your solemn assemblies.

22 Even though you offer me your burnt offerings and grain offerings, I will not accept them; and the offerings of well-being of your fatted animals I will not look upon.

23 Take away from me the noise of your songs; I will not listen to the melody of your harps.

24 But let justice roll down like waters, and righteousness like an ever-flowing stream.

25 Did you bring to me sacrifices and offerings in the forty years in the wilderness, O house of Israel?

26 You shall take up Sakkuth your king, and Kaiwan your star-god, your images, which you made for yourselves;

27 therefore I will take you into exile beyond Damascus, says the LORD, whose name is the God of hosts.

The practice of sacrifice and celebration named here belonged completely in the area of agricultural changes, and thus it was related to the structure of time. In that Amos criticized this practice, he indirectly turned against the structure of time connected with it. He did it as he reminded them of their own history:

1 Hear this word that the LORD has spoken against you, O people of Israel, against the whole family that I brought up out of the land of Egypt:

2 You only have I known of all the families of the earth; therefore I will punish you for all your iniquities. (Amos 3:1-2 [NRSV])

Translated into our terminology, in memory Amos called forth the historical dynamics of an asynchronism between transcendental and endogenous time. The Israelites, however, ignored Yahweh's interventions ("yet you did not return to me," Amos 4:8 [NRSV]) on the basis of their accommodation to the transcendent pole. Thus, the next step consisted of calling the Israelites back to this pole. Amos 5:4-5 [NRSV]:

4 For thus says the LORD to the house of Israel: Seek me and live;

5 but do not seek Bethel, and do not enter into Gilgal or cross over to

Beersheba; for Gilgal shall surely go into exile, and Bethel shall come to nothing.

At the end, the already mentioned criticism of sacrifice and cult followed, in which the disintegration of the dynamic tri-polar system took concrete form by regression to myth. At the same time, this criticism was connected with two constructive activities that restored the dynamics of the tri-polar system. For one thing, Amos urged the constructive formation of social time in the form of righteousness and justice, in place of the collective regression into bacchanalian giddiness (5:24). For another thing, Yahweh brought into play the effectiveness of God's transcendent pole, in that God announced a new action in history, even if by a negative sign, namely, the Exile "beyond Damascus" (5:27). Entering the Exile should decisively mold anew the understanding of time.

Before entering the Exile and following the preexilic prophets, the Deuteronomistic preachers (Wolff 1969, 132ff.) reinforced the prophetic impulses. They made demands for the centralization of the cult (Deuteronomy 12), and they reinforced the inculcation of the uniqueness of Yahweh (in the *sh^ema Israel*, Deut. 6:4: "Hear, O Israel: The LORD our God, the LORD is one" [NRSV, footnote n]), once again reinforcing the transcendent pole. Furthermore, at the same time they referred to the present moment of today. The accompanying asynchronism between transcendent and endogenous time offered the necessary dynamics for our tri-polar system. The word *hayyom* "today" occurs seventy times in this sense in Deuteronomy alone.

The memory of the past, the reference to the present, the holding onto the reference to God's transcendence — all these guaranteed to us the *openness of the time* of Yahweh.

> Know therefore that the LORD your God is God, the faithful God who maintains covenant loyalty with those who love him and keep his commandments, to a thousand generations, and who repays in their own person those who reject him. . . . Therefore, observe diligently the commandment . . . that I am commanding you today. (Deut. 7:9-11 [NRSV])

The experience of the Exile put an end to the desperate attempt of the prophets to visualize a return of the people to God based on their own ability, and it produced a new experience of time.

The Postexilic Experience of Time with
Second Isaiah and the Deuteronomists

The Babylonian exile was a crucial experience for Israel, just as were the Exodus from Egypt and the wilderness wandering. Their powerlessness in being endangered ("All people are grass, their constancy is like the flower of the field," Isa. 40:6b [NRSV]) and their being cut off from their own history in a foreign country ("My way is hidden from the LORD, and my right is disregarded by my God," Isa. 40:27b [NRSV]) were in conspicuous contrast to the victorious power of Marduk, the Babylonian god. Thus, the damage to their history represented a special danger for endogenous time. They could be pulled into the mythic-cyclic time of the Babylonian astral god, Marduk, and into its cultic observances. In this dangerous situation, a prophet arose from the people of Israel in exile. Due to a lack of knowledge of his real name, he has been given the name "Second Isaiah," and his writings have been passed on to us in chapters 40-55 of the book of Isaiah.

In facing the danger to endogenous time, how did Second Isaiah respond to the danger from mythic regression? Second Isaiah confronted the endangering of endogenous time with an extreme emphasis on transcendent time. For the first time, here in the Exile, Israelite monotheism managed a complete breakthrough. From the mouth of the prophet came many affirmations of the divine sovereignty ("I am God, and there is no other," Isa. 46:9b [NRSV]; 45:21b, 22 b). God alone was entitled to eternity, in contrast to God's creatures. With Second Isaiah we find the unique name "eternal God" (*'elohe 'olam Yahweh,* Isa. 40:28). God's eternity embraced all further times ("I am the first and I am the last; besides me there is no god," Isa. 44:6b). However, Second Isaiah did not stop here; he connected this divine sovereignty with God's activity as Creator:

> Have you not known? Have you not heard? The LORD is the everlasting God, the Creator of the ends of the earth. He does not faint or grow weary; his understanding is unsearchable. (Isa. 40:28 [NRSV])

However, now this creative activity of Yahweh included time! Yet, Yahweh did not create time in an abstract sense. In accordance with the Hebrew connection of time and event already familiar to us, God had the ability of foreknowledge and the ability to accomplish a desired event.

Who is like me? Let them proclaim it, let them declare and set it forth before me. Who has announced from of old the things to come? Let them tell us what is yet to be. (Isa. 44:7 [NRSV])

... declaring the end from the beginning, and from ancient times things not yet done, saying "My purpose shall stand, and I will fulfill my intention." (Isa. 46:10 [NRSV])

The text of Isaiah 44:7 is interesting for it establishes for the first time the terminology of the future *(otioth)*, as well as in Isaiah 41:22 *(habaot)*. That is, the future — as a future planned by God — became an autonomous category. Henceforth it was the interest of the exiled, especially since they had been cut off from their past. However, this future was not of interest for its own sake, but rather because Yahweh will do something new. God will create a new historical awakening for the exiles.

See, the former things have come to pass, and new things I now declare; before they spring forth, I tell you of them. (Isa. 42:9 [NRSV])

You have heard; now see all this; and will you not declare it? From this time forward I make you hear new things, hidden things that you have not known. (Isa. 48:6 [NRSV])

In this way the creative activity of Yahweh was connected with two new categories of "the future" and "the new." Both categories defined God's transcendent time over against repetitive cyclic-mythic time, and also over against the deterministic, astral environment of the exiled Israelites. The intention of Yahweh was to include Israel in new events by a new "everlasting covenant" *(berith 'olam,* Isa. 55:3; everlasting "salvation, deliverance, love," Isa. 51:6, 8; 54:8 [NRSV]).

Do not remember the former things, or consider the things of old. For I am about to do a new thing; now it springs forth, do you not perceive it? (Isa. 43:18-19 [NRSV])

Summarizing, we can say that the renewed danger of mythic regression of endogenous time was countered by Second Isaiah with his emphasis on transcendent time, with a newly expanded openness of time. Apparently, this continuous battle led to the emancipation from natural-cyclic time. With that emancipation, from the time of the Exile Israel developed a kind of linear idea of time, even if it was not in the sense of an abstract linear time that was free of events. At the same time, with the hope of a new beginning in the history of Israel that will be caused by God, reflection on the reasons for the historical catastrophe of the Exile began. The result of this reflection was the formation of the so-called Deuteronomic History, which claimed that Israel's failure (as claimed by the preexilic prophecy) was in their falling away from God; that is, it was seen as transcendent time.

Wisdom

The approach to life, and the formation of life, which was described as wisdom, was a phenomenon of all ancient oriental cultures, and it was not just limited to Israel. In this respect, it is clear that it dealt with human life in general, and that it missed out on the peculiarity of the transcendent time of Israel. A *stabilizing* transcendent time entered in place of a *dynamizing* transcendent time. God was not the creator of the new, but rather God was the preserver of what existed: "That which is, already has been; that which is to be, already is; and God seeks out what has gone by" (Eccles. 3:15 [NRSV]). In the sense of our tri-polar system, *de facto* from the start we have to do with a reduction of the two poles of endogenous and exogenous time (natural time and social time). Again and again we encounter this reduction, in which the dynamizing of the tri-polar system was canceled by the independently effective transcendent time.

Now, we can say that wisdom consisted in listening for order in nature and in human life, and in deriving from this listening maxims about behavior and action. This orderly thinking referred to various phenomena of life, such as health and illness, poverty and wealth, good and evil, and also time. As it was aimed at this ordering, wisdom sought to minimize and to tame the area of the opaque, the fateful, and therefore the contingency of events. In terms of our system it intended to produce a synchronism between endogenous and exogenous time in order to bring about the dependability and the stability of life. For this reason it is not remarkable that the area of contingent newness, history, played no role. However, again and again in our previous observations of Israel we had determined the dominance of transcendent time. Typically, the thinking of wisdom was thinking of order. The wise seek order in order to gain time. Wisdom was primarily encountered in the Old Testament books of Job, Proverbs, Sirach (in the Apocrypha), and Ecclesiastes.

Since wisdom replaced the active transcendent pole, it is not surprising that the reduction of the specific features of endogenous time entered particularly into consciousness. This reduction was found for the first time in the painful realization of human finiteness. We read in Job 14:1-2 [NRSV]:

> A mortal, born of woman,
> few of days and full of trouble,
> comes up like a flower and withers,
> flees like a shadow and does not last.

And even more drastically, Ecclesiastes notes: "For the fate of humans and the fate of animals is the same; as one dies, so dies the other. They all have the same breath, and humans have no advantage over the animals; for all is vanity" (Eccles. 3:19 [NRSV]). Here was the painful realization of the restrictions on human temporality, but also there was a mention of the danger of the inner disorder of endogenous time. The danger was present if humans did not find a synchronism with the order of exogenous time by perceptions and action. This order of natural exogenous time consisted in the recognizable periodicity of natural courses. In the area of social exogenous time, the causal order of actions and results could fit the manner of action. Therefore, one spoke of the wisdom of the connection between action and result. In social exogenous time, this unambiguous causal connection between actions and results should enable a person acting wisely to reach a successful formation of life by corresponding behavior.

> The wicked are overthrown, and are no more,
> but the house of the righteous will stand. (Prov. 12:7 [NRSV])

> Be assured, the wicked will not go unpunished,
> but those who are righteous will escape. (Prov. 11:21 [NRSV])

In addition, the wise were of the opinion that these arrangements had been made by God who directed them for the good of humans. Illness, evil, and guilt were therefore the consequences of human mistakes in this order. With piquant ambiguity for modern ears, we read: "He who sins against his Maker, may he fall into the hands of the physician" (Sir. 38:15 [NRSV, footnote y]). It is by no means the case that these orders for recognition and action were openly revealed. Rather, the art of wisdom consisted precisely in waiting for the "right time" for action, somehow. However, here was also the constant source of irritation for the wise. It is threatened by the spirit that lurked in skepticism, in elegant resignation, or in despairing hedonism. Wisdom marked a special vigilance for the course of temporal performances. The best-known witness of the perception of time by wisdom has been handed down to us in Ecclesiastes 3:1-8 ("For everything there is a season . . ."; see above, pp. 58-59).

However, that list concluded with the insight (already mentioned) about the difficulty, if not impossibility, for all to recognize the right time to act and to get ready for it ("What gain have the workers from their toil?" Eccles. 3:9 [NRSV]). That is to say, the synchronism that was sought between endogenous and exogenous time represented an ideal condition, which was postulated as an attainable goal. This view was asserted not least for the reason that God imposed an unambiguous limit of perceptibility on the given order of time. Humans could not fathom the secrets of time ("yet they cannot find out what God has done from the beginning to the end," Eccles. 3:11b [NRSV]), but they should make do with the feeling of God's superiority to time (eternity) ("He has made everything suitable for its time; moreover he has put a sense of past and future into their minds," Eccles. 3:11a [NRSV]).

The experience of time by wisdom, therefore, was also called a border experience. This border experience was imposed on sages, when they realized that the connection of action and result did not always work in the ethical area. Experience also taught that it was not rare for success to be denied to the devout, and that wickedness often triumphed. These two border experiences put in question the concept of the wisdom of life. As a way out of this dilemma, two solutions were offered: apocalypticism and the idea of righteous suffering. In both formations of life, a specific structure of time was expressed once again.

Apocalypticism

As with wisdom, so also apocalypticism originated as a religious phenomenon in all ancient oriental cultures. It is not surprising, since we have given reasons for the two grounds from which apocalyptic religiosity and wisdom were derived.[1] As to the time of the formation of the apocalyptic literature, we would assume it was about the fourth century before Christ in the Jewish Diaspora. From there the attitude of wisdom to life then extended in Jewish origins from Palestine to the intellectual leaders of the temple in Jerusalem itself. This manner of thinking was opened up there, and it was extended perhaps even to the sect in Qumran by the teacher of righteousness. The oldest sources in the Old Testament are the books of Daniel, and in the Apocrypha, the book of Tobit. Outside the canon there was a proliferation of apocalyptic literature, such as the Ethiopian book of Henoch, the Syrian book of Baruch, and so forth. Generally, these books arose in a time when the Jewish people were excluded from an active participation in their story, whether it was because of foreign rule, or because they were in the Diaspora. At the same time, this observation provides an indication of the application of our heuristic principle for the interpretation of the peculiarities of apocalyptic literature.

The experience of historical powerlessness and the exclusion from independent political formation led to the wisdom literature, and it increased the weakness of transcendent time as a purely preserving quantity. Among other things, an apocalyptic God appeared to be banished into a completely unattainable transcendence. Also, God's function was weakened as guarantor of the ethical order for the connection of action and result. That happened because apocalyptic literature was certain that God gave room for evil, although it was restricted with regard to time, as already the wisdom of the book of Job had indicated. It followed that evil could extend itself freely, and it accumulated. A picture of history originated as a process of continual disintegration. This social and cosmic process of disintegration had a temporal order throughout, and it had been determined by God. It appeared in the form of the so-called doctrine of two aeons. This doctrine presented a trans-historical end of history. After an accumulation of catastrophes, God would usher in the end in a final battle on a cosmic scale (Armageddon) between good and evil.

---

1. This derivation is not indisputable. Primarily since the Second World War, the thesis has been accepted that Jewish apocalypticism derived from prophecy. Gerhard von Rad's thesis in his *Theology of the Old Testament* of an inner affinity between wisdom and apocalypticism did not gain acceptance. Today the scholarly world is divided on this question, and thus nothing will be added because there is no generally accepted conception of apocalypticism (cf. Lebram 1978, 192-99).

All these aspects were shown quite clearly: transcendent time was no longer involved in the current course of history, and its function was greatly reduced to the conservation of order. It was simply involved at the beginning of history when it put in place the deterministic course of history, and it will be involved at the trans-historic end of history. Humans have been placed in between the times. The time of salvation was only a mythological beginning and ending of history. Apocalypticism swung back and forth between mythological archaism and futuristic fantasy. The present was not actively formed either by the hope of renewal through wise action, or by the transcendent time of God. In fact, this cutting off of humans from transcendent time showed up in many ways in their actions in the present. We have already seen that action and the active formation of life were constitutive for the Hebraic perception of time.

Yet, in apocalypticism this aspect of action led back to contemplation and to a pure rationality that observed and schematized. Its products (Lebram 1978, 192) were anonymous and pseudonymous works of the respective apocalyptic authors. They consisted of dreams and visions as sources of revelation, and there were cosmic speculations that included interpretations of astrology, as well as allegory and numerology. The game of switching between action and rationality as part of the formation of endogenous time had collapsed in the cancellation of action. For endogenous time, at best there remained only a formless interpretation of events, and at worst a schematizing rationality. This rationality in connection with its affiliated linear time — by its unwavering form of history that was secure in the confrontation with mythology — now indulged in accumulating and compiling knowledge from books that were unproductive. Linear time could be time that was absolutely open, in which event and time were associated with one another. However, time in apocalyptic writings became an uneventful route, as in the schematizing use of the word ʿolam (eras of the world) (Jenni and Westermann 1979, especially 228ff.). The same view of time was expressed in the deterministic order of time established by God. Apocalypticism received all of this by secret revelations in reveries and visions of God.

In this way, apocalypticism replaced action in time with knowledge about time. The apocalyptic writer did not form his time, he suffered it, even if glossing over it rationalistically and intellectually. The apocalyptic writer turned into a dependent being. This dependence appeared in their high evaluation of exogenous natural time, and that was completely atypical for earlier Israel. The lunar calendar gained importance again. Astrology and all sorts of cosmic speculations sprang up.

In the sense of our tri-polar system, we can interpret apocalypticism as a most extreme product of disintegration. Transcendent time was unable to participate actively in giving impulses for maintaining the openness of time in a dynamic tri-polar system. The result was that a reduction took place to the level of the bipolarity of endogenous and exogenous time. Furthermore, this system lacked any dynamics, because in the absence of human action the attempts at synchronization of endogenous time and exogenous natural time took the form of astrology and cosmic speculations. The parts of the system were isolated from each other, and therefore the system collapsed. In the end there is an individual piety that had spun a fantasy for itself, as a silkworm goes into a cocoon. Here, an impulse from transcendent time could only help, and now we stand at the threshold of the New Testament.

NEW TESTAMENT

Jesus: The Message of the Approaching Kingdom of God

The preaching of Jesus began with the statement: "Repent, for the kingdom of heaven has come near" (Matt. 4:17 [NRSV]). In the sense of the old prophetic and messianic tradition of Israel, it was clear that the transcendence of God was announced as breaking into the present from the future. God's kingdom was expected and understood by apocalyptic writers as the end of everything. Here it was the beginning of everything, and it was the starting point for actively reversing thinking, feeling, and acting. With Jesus the tri-polar structure of time appeared, as it were, "forward," "tilted" toward the future:

Exogenous time
God's kingdom with reference to the world

                                           Transcendent time
                                           God's coming kingdom

Endogenous time
God's present kingdom

Let us look at the three poles of the structure of time in Jesus' message concerning the approaching kingdom of God:

a. Transcendent Time

The experience of transcendent time as the time of the coming kingdom of God came to light particularly in Jesus' parables. The coming kingdom God can only be spoken of symbolically since it is not directly available. It was important that the hidden advent of the kingdom of God was included in the "here and now" of Jesus' present. The full lordship of God transcended the future, but it was "close at hand." In parables of seed and harvest, Jesus clarified this connection. It was generally presented in the parable of the mustard seed:

> He also said, "With what can we compare the kingdom of God, or what parable will we use for it? It is like a mustard seed, which, when sown upon the ground, is the smallest of all the seeds on earth; yet when it is sown it grows up and becomes the greatest of all shrubs, and puts forth large branches, so that the birds of the air can make nests in its shade." (Mark 4:30-32 [NRSV])

Here hope was awakened: as small as the mustard seed is, still the mustard shrub is big if it grows. It is so big that birds find refuge in its shade — a picture of security and peace! So it was with the beginnings and origins of the lordship of God: it was a harbinger of a new and redeemed world, in which there will be peace. They are sparks from the great light, which was approaching from the eternal future of God to liberate the world. Jesus said: "But if it is by the finger of God that I cast out the demons, then the kingdom of God has come to you" (Luke 11:20 [NRSV]).

Transcendent time was the dimension of God's inexhaustible possibilities, which were allowed to flow in from the future into the world in order to renew it. Transcendent time was therefore understood as the source of all endogenous and exogenous time. This message of Jesus concerning the approaching kingdom of God dispensed with the nervousness of the apocalyptic writers. It radiated not only hope, but also peace. It became especially clear in the parable of the growing seed:

> He also said, "The kingdom of God is as if someone would scatter seed on the ground, and would sleep and rise night and day, and the seed

would sprout and grow, he does not know how. The earth produces of itself, first the stalk, then the head, then the full grain in the head. But when the grain is ripe, at once he goes in with his sickle, because the harvest has come." (Mark 4:26-29 [NRSV])

As the earth produces fruit "of itself," so the lordship of God is quietly and "of itself" placed where the ground is fertile. Transcendent time is "really" transcendent in the sense that it is completely taken away from human desires, plans, and actions. To put it in philosophical terms: for Jesus, transcendent time represents the possibility of experiential time. Endogenous and exogenous time grows "of itself" from transcendent time.

b. Endogenous Time

Since God's kingdom was already present in Jesus and his personality, transcendent time changed into an especially qualified endogenous time. Jesus said, "The kingdom of God is among you" (Luke 17:21 [NRSV]), and he indicated by this that we find God's time in ourselves as an inner time, which is free of the forces of exogenous time. It is "fulfilled time," which permeates us with feelings of trust, gratitude, and salvation. That became especially clear in Jesus' "call of the savior":

"Come to me, all you that are weary and are carrying heavy burdens, and I will give you rest. Take my yoke upon you, and learn from me; for I am gentle and humble in heart, and you will find rest for your souls. For my yoke is easy, and my burden is light." (Matt. 11:28-30 [NRSV])

The phrase "rest for your souls" points more to the inner realm of endogenous time, which enables self-discovery and self-becoming — independent of social and natural providers of time. This protected and autonomous inner space of our own time belongs entirely to us because it belongs entirely to God. It breaks through the unending chain (arrested by the concept of linear time) of care and anxiety about the future. Therefore, in the Sermon on the Mount we find Jesus' call to renounce worries: "Therefore do not worry, saying 'What will we eat?' or 'What will we drink?' or 'What will we wear?' For it is the Gentiles who strive for all these things; and indeed your heavenly Father knows that you need all these things. But strive first for the kingdom of God and his righteousness, and all these

153

things will be given to you as well. So do not worry about tomorrow, for tomorrow will bring worries of its own. Today's trouble is enough for today" (Matt. 6:31-34 [NRSV]).

Endogenous time was perceptible here as the fulfilled present that sets us free from unnecessary worries and fears. It is established in trusting God's possibilities, which come toward us from God's eternal future each day.

c. Exogenous Time

Jesus' preaching did not stop with the endogenous time of fulfilled time. It was not about the redemption of the individual soul, but it was about the redemption of the earth. Endogenous time changed into exogenous time, not the reverse, as in the mythic religions. Again and again, Jesus sharpened the reference of God's kingdom to the world, and he reminded us of the social responsibility of individuals, as, for example, in the parable of the day of judgment (Matthew 25), or in the story of the good Samaritan (Luke 10). He admonished people to interpret the "the present time" (Luke 12:56). He even broke through the rigid pattern of exogenous time, when he healed on the Sabbath (Luke 13:10-17). In the hearing and doing of his words, a "new righteousness" was created (Matt. 5:20), which provided a new definition of exogenous time: it was a time in which love and social responsibility should be realized. However, it does not happen independently, but rather it always happens in view of transcendent time as enabling my life and action.

It is clear: For Jesus the tri-polar structure of transcendent, endogenous, and exogenous time is located in a dynamic balance, which is open to the future. This balance was aimed at the liberation and salvation of humans and the whole creation.

Paul: The Righteousness of God
and the Turning Point of History

The concept of the "kingdom of God" no longer played a central role in the preaching of the apostle Paul. Instead, "the righteousness of God" entered as the center of his message. Why? As E. Jüngel (1972) has shown, there was no lack of agreement with Jesus' message of the approaching kingdom of God, but rather there was a change of perspective with regard to time: Jesus looked forward to the advent of the messianic kingdom as a future event. Paul looked back on the advent of the messianic kingdom as a final event: in the cross and resurrection of Jesus the world was reconciled to God and the turning point in history has already happened! What Jesus hoped and begged for has happened: the lordship of God as the renewal of the world has already happened in the resurrection of Jesus from the dead, at a certain place and at a certain time in the reality of history. Therefore, the concepts shifted: where Jesus spoke of the "kingdom of God," Paul talks of the "righteousness of God." In the central atoning events of the cross and the resurrection, God's justice was aimed at us — to bring us salvation. The tri-polar structure of time in the proclamation of the apostle can therefore be sketched as follows:

Old aeon	Advent of the new aeon in the old	New Aeon
Exogenous time The righteousness of God in the everyday life in the world → LOVE		

- -

	Transcendent time (arrived) The revelation of the righteousness of God in Christ → ORIGIN OF EVERYTHING NEW	*Transcendent time (coming)* The universal goal of the righteousnss of God → HOPE

- -

Endogenous time
The righteousness of
God that takes me →
FAITH

a. Arrived Transcendent Time

Paul writes in the letter to the Galatians:

> But when the fullness of time had come, God sent his Son, born of a woman, born under the law, in order to redeem those who were under the law, so that we might receive adoption as children. (Gal. 4:4-5 [NRSV])

The new aeon, which apocalyptic writers expected in the distant or near future, has already broken into our midst: in the man Jesus of Nazareth. The invasion of this transcendence into history is unique and unrepeatable, and it is quintessentially the irreversible event!

God has forced open the world of the old aeon, which Paul described as a prison, in which the law, sin, and death prevailed. Paul had his own existential experience of the invasion of the new aeon into the old aeon in the instant of his conversion. He looked back gratefully on this "transcendent" event (Gal. 1:12, 16). The time of the world and his own lifetime were qualified for Paul by this influx of transcendent time in an irreversible manner. God's revelation in Christ is the decisive basis for structuring our own endogenous time in a new way.

b. Endogenous Time

Paul describes in Romans 5 how the inner experience of time was changed by faith in Christ:

> Therefore, since we are justified by faith, we have peace with God through our Lord Jesus Christ, through whom we have obtained access to this grace in which we stand; and we boast in our hope of sharing the glory of God. And not only that, but we also boast in our sufferings, knowing that suffering produces endurance, and endurance produces hope, and hope does not disappoint us, because God's love has been poured into our hearts through the Holy Spirit that has been given to us. (Rom. 5:1-5 [NRSV])

Paul went even a step further in the Letter to the Galatians. He wrote:

For through the law I died to the law, so that I might live to God. I have been crucified with Christ; and it is no longer I who live, but it is Christ who lives in me. (Gal. 2:19-20 [NRSV])

In faith we receive our lives and also a new center for our experience of time: Christ himself, who lives in us. He reigns, and he also rules and controls our perception of time. How differently would Paul have experienced his many distresses (cf. 2 Corinthians 11) without the presence of Christ in him, that is, without endogenous time filled by him!

c. Exogenous Time

In Romans 12, Paul clarified how the Christian experience of endogenous time was related to the experience of exogenous time:

I appeal to you therefore, brothers and sisters, by the mercies of God, to present your bodies as a living sacrifice, holy and acceptable to God, which is your spiritual worship. Do not be conformed to this world, but be transformed by the renewing of your minds, so that you may discern what is the will of God — what is good and acceptable and perfect. (Rom. 12:1-2 [NRSV])

Paul pleaded for a specifically Christian asynchronism of endogenous and exogenous time: It all depends on whether we are formed by the clocks of this world, or whether we undertake responsibility in the world by the renewal of our minds. Exogenous time does not determine endogenous time. It is the other way around. It follows that the "spiritual worship," which lets the righteousness of God come into its own in the everyday life of the world, is for the honor of God and for the benefit of people and their world.

d. The Coming Transcendent Time

Undeniably, for Paul there is an "eschatological proviso" remaining in the tri-polar structure of time: "For now we see in a mirror, dimly, but then we will see face to face. Now I know only in part; then I will know fully, even as I have been fully known" (1 Cor. 13:12 [NRSV]). Paul was not a Gnostic or a mystic, who experienced transcendent time as an eternal present. He hoped! Two hundred and fifty years later, however, the Neo-platonist Plotinus, who wanted to escape from the world, praised the overcoming of hope as the highest form of inspiration! An alert look at the unredeemed world and unredeemed nature was enough for Paul to recognize the enslavement of this world as a consequence of the old aeon. On the contrary, he hoped with all his strength for the coming aeon of the total freedom of all the creatures in God. In Romans 8, this hope is formulated pregnantly:

> I consider that the sufferings of this present time are not worth comparing with the glory about to be revealed to us. For the creation waits with eager longing for the revealing of the children of God; for the creation was subjected to futility, not of its own will but by the will of the one who subjected it, in hope that the creation itself will be set free from its bondage to decay and will obtain the freedom of the glory of the children of God. We know that the whole creation has been groaning in labor pains until now; and not only the creation, but we ourselves, who have the first fruits of the Spirit, groan inwardly while we wait for adoption, the redemption of our bodies. For in hope we were saved. Now hope that is seen is not hope. For who hopes for what is seen? But if we hope for what we do not see, we wait for it with patience. (Rom. 8:18-25 [NRSV])

It is clear that with Paul the tri-polar structure of time found the same dynamic balance that it had with Jesus. Time is experienced and thought of as an open system, which is opened by and exists from the coming God. The experience of time connected with it can rightly be called "messianic" (Moltmann 1985, 134). It is the experience of a "new creation" (2 Cor. 5:17) in the midst of the old world, which by this "new" has become irrevocably obsolete. For the future we find open structures of modes of time that gain a new meaning in the light of the advent of this Messiah (Moltmann 1985):

> All turns into the past, which in the present of the Messiah Jesus is no longer valid and no longer works: sin, "the law," and death.
> All turns into the present, which in the present of the Messiah Jesus is already valid and works: grace, reconciliation, freedom.
> All turns into the future, which is now experienced as not yet, already as hope: resurrection of the dead, redemption of the body, and eternal life.

John: Biblical Mysticism

In the Gospel according to John and in the First Letter of John — besides the Synoptic Gospels and the letters of Paul — there is another important branch of the tradition of the New Testament. We can examine this branch for its theological understanding of time. A structure appears that is comparable with Paul's:

Exogenous time
"In the world you face persecution.
But take courage!"
(John 16:33 [NRSV])

Transcendent time (arrived)	*Transcendent time (coming)*
"And the Word became flesh and lived among us." (John 1:14 [NRSV])	"What we will be has not yet been revealed." (1 John 3:2 [NRSV])

Endogenous time
"Those who abide in me and I in them
bear much fruit." (John 15:5 [NRSV])

a. Arrived Transcendent Time

Already in the prologue of the Gospel according to John (John 1), all the important motifs of this Gospel were mentioned, particularly the motif of the eternal Logos who descended and was incarnate: "And the Word became flesh and lived among us, and we have seen his glory, the glory as of a father's only son, full of grace and truth" (John 1:14 [NRSV]). The evangelist looked back to the unique and irreversible events of the incarnation of the eternal word in time. The central theme of his message was the temporality of transcendent time in Jesus' life, particularly in his suffering, dying, and rising. The invasion of transcendent time into the time of the world appeared as the Son's honoring of the Father in the Holy Spirit. In Jesus' death on the cross, this work reached its goal. The Johannine Christ said at the cross: "It is finished" (John 19:30 [NRSV]). Pending transcendent time was not diminished in its meaning, but rather the finality of the redemption through Christ was emphasized. Those who are chosen and redeemed have a share in the temporality of the eternal on earth since it happened in Christ: "From his fullness we have all received, grace upon grace" (John 1:16 [NRSV]). Christians' inner experience of time has changed.

b. Endogenous Time

The inner experience of time for redeemed humans, according to John's Gospel, was characterized by the concept of "bearing fruit," or to say it in modern terms, creativity:

> Abide in me as I abide in you. Just as the branch cannot bear fruit by itself unless it abides in the vine, neither can you unless you abide in me. I am the vine, you are the branches. Those who abide in me and I in them bear much fruit, because apart from me you can do nothing. (John 15:4-5 [NRSV])

It becomes clear in the picture of bearing fruit that the life of Christians was determined and guided not by outer dimensions and rhythms of time, but rather by the inner dimension of what was "ripe" and "at the time." Jesus himself pointed his disciples toward this endogenous dimension when he talked again and again about whether or not "the hour has come" in order to accomplish the revelation (cf. John 7:6; 17:1).

In Jesus' discipleship, the disciples should learn to wait for the "right hour" and to bear fruit when the time is fulfilled. "Abide in my love" (John 15:9 [NRSV]); that provided the possibility of bearing good fruit. It has been called Johannine "mysticism" (Theißen 1994, 147-63), in which God not only sees and perceives us from above, but also abides in us. If we abide in love, we are in God and God is in us, just as the Son is in the Father and the Father is in the Son. This mysticism of love became especially clear in 1 John 4:16: "God is love, and those who abide in love abide in God, and God abides in them."

c. Exogenous Time

The Johannine writings clearly had a more reserved and a more detached relation to external time and to the time of the world of the old aeon than Paul or Jesus. The First Letter of John said:

Do not love the world or the things of the world. The love of the Father is not in those who love the world; for all that is in the world — the desire of the flesh, the desire of the eyes, the pride in riches — come not from the Father but from the world. And the world and its desire are passing away, but those who do the will of God live forever. (1 John 2:15-17 [NRSV])

Nevertheless, there was a positive counter-reaction of the experience of endogenous to exogenous time: it was the courage and bravery of Christians in the everyday life of the world. The Johannine Christ said in his farewell discourse: "In the world you face persecution. But take courage; I have conquered the world!" (John 16:33 [NRSV]). The world's external time that fades away was experienced as depressing and pressing, as we speak today of "the pressure of time" and "schedule pressures" and "the harshness of life." It is valid, however, to resist this pressure through the victory of Christ: once and for all, he has conquered the world, its rush and its pressure in the cross.

d. The Coming Transcendent Time

Christ conquered the world, but this victory was not yet public and really accepted all over the world. Therefore, openness to the future of the coming world remained an important pole in the tri-polar structure of time. The First Letter of John said:

Beloved, we are God's children now; what we will be has not yet been revealed. What we do know is this: when he is revealed, we will be like him, for we will see him as he is. (1 John 3:2 [NRSV])

The final truth of the world, therefore, was still not apparent, and the coming transcendent time will bring it to light.

To summarize, there was a certain proximity in the Johannine literature to the type of "mystic experience of time," as we described it. Nevertheless, with a more exact comparison of texts it turned out that here time was understood as an open system, which was stamped as tri-polar by the three dimensions of faith, love, and hope.

Discipleship: Renewal of Time by Self-Denial?

On the one hand, in the three preceding chapters, we established that, in the nearness of the kingdom of God, transcendent time was *primarily* effective in Jesus' life (which corresponded with our beginning to speak of the *effects* of transcendent time). On the other hand, the fullness of its realization was still pending. In addition, we saw in qualitative considerations that these effects *secondarily* changed believers. It is now time for us consider how to specify in more detail these effects on our three anthropological structures of the experience of time.

If the new time of God's kingdom has already interrupted the old time, but did not arise from it, then it clearly confirms the structure of the time of God's kingdom. Furthermore, it confirms the threefold structure of the time of humans, but at the same time it is open to a more extensive dimension. We now want to discuss this relationship of confirmation and expansion.

- First of all, it is clear that the mythic-cyclic structure of time in human experience can hardly be brought into harmony with the structure of time of God's kingdom. Such is the case with Jesus, and also with Paul, for they spoke unequivocally of a certain devaluation of the basic biological structures of the human experience of time, since the latter belong, so to speak, to the old Adam. Rather, in consideration of the form of time of God's kingdom, Jesus and Paul broke through the biological control circuit of the mythic-cyclic form of time and its sluggishness. Jesus told his disciples in the garden of Gethsemane: "Keep awake and pray that you may not come into the time of trial; the spirit indeed is willing, but the flesh is weak" (Mark 14:38 [NRSV]). The linear understanding of time came closer to portraying the effect of the kingdom of God in somewhat more detail, in so far as one can conceive of it, by drawing a linear time line.
- Finally, the examination of John and Paul showed that the mystic structure of time has spoken to humans.

With a certain restriction concerning the cyclic structure of time, all three forms of the endogenous experience of time were received positively. The affirmation of the basic structures of time in our human existence is not to say, however, that they exhaust the acts of God, as we already indicated in the "eschatological reservation." What is meant by the greater abundance of the structure of time that is still pending for the participants in God's kingdom, if this structure is indeed announced in believers, even if only in passing and embryonically? To say it briefly, what *temporal advantage* do believers have, as opposed to nonbelievers? What richer structure of time do believers have, as opposed to the threefold structure of time already realized by humans who are living full lives? And, how can one reasonably tell humans that, besides this threefold structure of time, God may give another more abundant structure of time? And how should one attain it?

If we get down to trying to answer these questions, then we must be clear that our threefold human structure of time is most deeply connected with the structure of our selves. We can connect the cyclic structure with the preconscious system of self, the linear with the mature and stable self of an adult, and the mystic with the opening to transcendence, as well as holistic experience.

When Jesus called humans to discipleship, and when for the sake of God's kingdom he appealed to us to ignore all others ("But strive first for the kingdom of God and his righteousness, and all these things will be given to you as well," Matt. 6:33 [NRSV]), then he put in question also the natural system of the selves of humans and their threefold structure of time. In fact, it is not only difficult for humans to conceive of it, but it is also an unreasonable demand. Therefore, already in the narratives of the Gospels, minds were divided about following Jesus (Matt. 10:34-39). Perhaps it is most obvious in Matthew in the narrative of the rich young man, whom Jesus wanted to call to discipleship (Matt. 19:16-26). By his flawless religious life, this young man had certainly attained a high degree of fulfillment of his humanity. However, he evidently could not correctly understand discipleship for the sake of God's kingdom. He remained stuck in the sluggishness of time in his own system of self, and he refused. There is another circumstance as well that makes the discipleship of Jesus appear less attractive at first. Jesus himself connected discipleship with suffering: "If any want to become my followers, let them deny themselves and take up their cross and follow me" (Matt. 16:24 [NRSV]).

In addition, Jesus led his disciples again and again into difficult earthly conflicts. Briefly said, the discipleship of Jesus put his supporters under considerable pressure. Again and again, disciples had to learn to get ready for changing situations, to deal with unexpected incidents, and to endure onerous situations. They did not have any clever recipe in the form of some laws, as the Pharisees had. Behind this lifestyle one can look for the strategy of letting the self mature through constant contact with the world, in order finally to overcome it, in favor of the new being of humans. One almost has the impression that Jesus systematically rehearsed with his disciples the psychological mechanisms for the tolerance of ambiguity, for the tolerance of frustration and the need for delay. This practice matured the system of the self, and also it allowed the structure of time connected with the psyche to mature — with the goal of making ripe the structure of time of God's kingdom. The pressure on the sluggishness of time of the three-fold inner systems of time of humans is in the service of life, even if only a few people realize it: "For the gate is narrow and the road is hard that leads to life, and there are few who find it" (Matt. 7:14 [NRSV]).

We break off our considerations here, because we approach the difficult theological problem of the relationship of nature and grace. Only a few hints are given here for the readers to do their own thinking along this line. It is apparent that the structure of time of believers and those justified by grace is really more abundant than those who at least want to realize their entire humanity. It is expressed by a quadruple openness:

- It is openly and perceptibly in contrast to the natural threefold structure of time.
- It is openly and perceptibly able to cope with the harshness of sophistication.
- It is openly directed to the coming God.
- It is openly compared with a mainly gracious reforming of the system of the self.

On four sides, the screws will be put to the "I" concerning renewal and overcoming. Paul, for whom the new time of God apparently became a reality of endogenous experience, is a witness for the renewal of time through self-denial, when he wrote: "And it is no longer I who live, but it is Christ who lives in me" (Gal. 2:20 [NRSV]). This denial of self is of course death — death of the old human — but at the same time the birth of a new one. The renewal of time through change is the horizon in which Jesus himself put the unreasonable demand for suffering discipleship: "Then Jesus told his disciples, 'If any want to become my followers, let them deny themselves and take up their cross and follow me. For those who want to save their life will lose it, and those who lose their life for my sake will find it'" (Matt. 16:24-25 [NRSV]).

SUMMARY

In the chapter "The Time of God," we showed that only in the Israelite-Jewish culture, in contrast to all other ancient oriental cultures, was there a real transcendence, and so there one can speak of a transcendent time. Transcendent time alone formed an autonomous element in our tri-polar structure of time. Concerning transcendent time *itself*, we did not want to make any statements, limiting ourselves to its *effects* with reference to humans.

In the course of Israelite religious history, this unique effect of transcendent time in Israel led to breaking the solidarity of the mythic-cyclic structure of time in favor of an open horizon of time. However, this openness of time had to be defended by the prophets (preexilic prophecy, Deuteronomic preachers) again and again from the danger of mythic regression, in order to protect the endangered continuity of history by remembering and hoping. In postexilic times monotheism became generally accepted, and, in view of the hopeless historical situation of the exiles, the horizon of history was brightened by the promise of a new historical action. There is a historical coincidence of the loss of home and space by the Israelites (Exile) with their orientation to the future in their religious alignment (hope for the Messiah, return in the old homeland, national resurgence to greatness, hope of new saving actions of Yahweh). However, in the context of the Israelites' being excluded from the current political exercises of power, the transcendent pole lost formative power, and the Israelites drew back into wisdom, and consequently into the dismal apocalyptic concepts of time. The solidification of time in Israel was manifested in apocalyptic anxiety (apocalyptic writers and Essenes), illusory political messianism (Zealots), the exemption from time by nomistic interpretation (Pharisees), and the adaptation to Greek metaphysical concepts of time (Sadducees, Philo of Alexandria).

Into this nebulous variety of concepts of time, Jesus of Nazareth brought more clarity with his announcement of the nearness of the kingdom of God in the form of his person. In his person there was tension between the lordship of God that had already begun as a sign and its pending completion. Transcendent time was connected with Jesus' immanent time, without surpassing it. It is clear that neither the indwelling of transcendent time in Jesus' immanent time nor its still pending completion in the three-fold structure of time of humans (mythic, linear, and mystic) can be expressed completely.

By faith, which grows in the discipleship of Christians in opposition to the world, humans can share in this process of the basic renewal of the systems of the self (grace) and gain a new structure of time. Joy and creativity are indications of sharing with Christ in the power of renewing time. We now want to try to formulate this new structure of time more exactly.

"The True Time"

SYSTEMATIC CONSIDERATIONS

INTERLACING THE MODES OF TIME AND OF FAITH

Georg Picht (1980, 362-74), A. M. Klaus Müller (1972), Arthur Prior (1967), Niklas Luhmann (1975, 103-33), and Reinhart Koselleck (1973) have demonstrated that the one-dimensional conception of time as a straight line, which leads linearly from the past to the present and future, is inadequate. The linear time of classical physics was a "prepared time" (A. M. Klaus Müller), that is, a reduction of the fullness of time to a straight line that had its meaningful function in the framework of classical mechanics. However, it did not adequately represent the reality of time. As Augustine had already clearly recognized in his meditations on time (Augustine: *Confessiones* 9, particularly 20, 26), it is absolutely necessary for the conceiving of true time to seek an interlacing of the modes of time: we can only explore and recognize the past if it has left traces in our present. It is only in the form of present documents that it lets itself be objectified. In the widest sense, what the historian sees is not the past itself, but the present of the past (PrPa).

Likewise, we know nothing about the future in itself — indeed, the essence of the future is exactly that it is not yet — and we can only make statements about things that can be done in the future, that is, what already is present in the form of probability and expectation. That is what each futurologist deals with, not with the future in itself but the present of the future (PrFu). What is left to be determined is this: what objectifying science uses, which is partly the present of the past (PrPa), partly the present of the future (PrFu), and accordingly also the present of the present (PrPr). Represented schematically:

PrPa ——————— PrPr ——————— PrFu

It is clear, however, that this interlacing series does not yet represent the complete structure of the modes of time. That is why it is necessary to represent the interlacing of all nine possibilities of past, present, and future in a kind of "matrix of time":

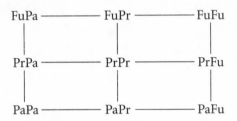

G. Liedke undertook the attempt (1974) to interpret this matrix, so that the middle horizontal row, PrPa/PrPr/PrFu, is the area of objectification. Therefore, it represents the area of science in the widest sense. The central vertical row, FuPr/PrPr/PaPr indicates the area of art and myth. Liedke (1974, 15-16) expounded:

> In the performance of the ritual of myth, for example, the Babylonian celebration of New Year's is not only experiencing the present of the new creation of the world, but also the first creation of the world as the future new creation. Humans were experiencing not only the past, but also the present and the future, because the mythic experience takes place simultaneously in the past, present, and future. It cannot therefore be objectified, because it cannot be reduced to what appears to us from the past and the future in the present. Likewise in the experience of art, particularly of music: the experience today of a symphony by Beethoven unites me with every past and future experience of this musical work, and therefore my present is present in the past and in the future. That is why we see that the series PrPa/PrPr/PrFu (science) and FuPr/PrPr/PaPr (art and myth) are not mutually exclusive. That is because each musical work is also a phenomenon of acoustics, only that, of course, a purely

physical and acoustic contemplation misses the artistry in the work of music, because it moves the past and the future of the musical experience into the present and reduces it to science.

Liedke (1974, 16) suspected that the four corners of the matrix (FuFu, PaFu, FuPa, and PaPa) cast light on time from the standpoint of faith:

> The interlacing of PaPa, PaFu, FuPa, FuFu does not contain the present, and therefore it does not contain the reality of phenomena. It describes the unavailable area of time, that which — if at all — is not directly experiential but at most contains the connection with the other interlacing with the Pr. Such indirect experience is the experience of faith.

Starting from this supposition, we can now say:

- Past of the past (PaPa) is, through God's revelation in Christ, the offense that was committed, and it belongs purely to the past in our faith: the power of the law, the power of sin, and the power of death. The biblical message of the forgiveness of sin, of the end of the law, and of the death of death is therefore reasonable as the message of the past of the past. In principle it escapes every scientific objectification, but it becomes accessible in faith.
- Future of the future (FuFu), therefore, is that which exceeds all extrapolation of the past into the future, the real "newness" that has not yet appeared. However, in faith it is the hope for eternal life, the new heaven and the new earth, in which God becomes all in all.
- Past of the future (PaFu) is that which has already appeared from the future in Jesus Christ. It is extracted from our present experience and realization. That is exactly the specifics of faith: the solid confidence is already there in all the final possibilities of the completion of the creation of the world, in the incarnation of God in Jesus Christ, and in the cross and resurrection Christ. The Christian faith refers to the past of the future as God's acts in the past. They already contain all the future acts of God.
- Finally, the future of the past (FuPa) is that of the past that is already of the future, that is, the presence of the past in the future.

This interlacing of the modes of time . . . marks the place that we call faith in the certainty of salvation. It lets us hope certainly that not only all future in the PaFu of the central events of salvation in Jesus Christ is already actually there, but also that the salvation that appeared in PaFu is identical with the future salvation being brought by God to us. That is, that the hope of faith can be without fear, because it knows that God's promises of the future with creation, incarnation, cross, and resurrection are true. (Liedke 1974, 17-18)

By following Liedke we can now say: the realization of the fullness of time in the interlacing of its modes gives us the desired theoretical clarity. Thus, we can discern the content of this objectifying knowledge of living faith in such a way that both knowledge and faith appear in their genuine brightness, and they become plausible.

TIME AS AN OPEN SYSTEM AND HOPE

As we have seen, the mythic, the rational-linear, and the mystic experiences of time presuppose that time is closed, that is, a symmetrical system. A branch of physics, thermodynamics — or, to say it more exactly, the second law of thermodynamics — compels us unavoidably to see that a basic difference exists between past and future (Weizsäcker 1971, 172-83) and that real time has an irreversible direction. It will even undertake the attempt to anchor the *direction* of time as a basic concept of scientific time and to formulate it mathematically, which is to the disadvantage of the concept of *symmetrical linear* time in classical mechanics. How this attempt will work out in the future cannot yet be seen (Prigogine 1970, 182-211; Prigogine and Stengers 1992). Besides the direction of time, its openness is also to be emphasized. Here degrees of openness can be distinguished.

With the closed systems of classical thermodynamics, the openness of time cannot be addressed because of the increase of entropy. On the contrary, because of the death by heat in closed systems, time runs up against a wall, so to speak. This situation changes only with open systems that can reduce their inward increase of entropy by an exchange of energy with their surroundings. Such systems succeed by maintenance of the status quo. However, only systems that are open, dynamic, and self-catalytic can be said to be self-organizing systems with an openness of time. Of course, the future of such a system is not predictable, but it is nevertheless determined. It is, so to speak, nailed down and restricted by the past, but with

170

unpredictability there is a relaxed openness of time. Such self-organizing feedback systems are described by chaos theory. The open structure of the time of faith, as we sought to illustrate it schematically in our matrix of time, is even richer in structure than the open structure of time in chaos theory. It is richer in structure in that the unpredictability of newness is not an extension of the past. It is really a contingent newness from the future or, more precisely, from God's advent.[1]

The philosophy of modern physics is on the way to a conception of time that we recovered in the central passages of the Bible (see above). For the Israelites freed from Egypt, for Israel's prophets, for Jesus, Paul, and John, there is the true time of history. It is a holy history that is in principle open to the future, and it is eaten from the future as from an invisible source.

In systematic theology this openness of time was often concealed in more or less subtle ways (Moltmann 1995, 22-39). It could be concealed by a "temporalizing of eschatology," which portrays the open future (FuFu) as a concept of linear time, as it appears probably most blatantly in the theology of time of Oscar Cullmann. Or, it could be by a "perpetuation of eschatology" (by Karl Barth, Paul Althaus, Rudolf Bultmann), in which God's transcendence from the future is twisted again into an (in the final analysis, as the Greeks understood it) "eternal present," which is "vertical" at every point of the axis of time. As Jürgen Moltmann showed (1995, 39-47), however, the Bible speaks consciously of the coming God (not of the "future God"), and of *newness*, which precisely does not "become obsolete" after its entrance into the present. In the biblical understanding, the "future" is not "future," but an "advent." Accordingly, systematic theology has to formulate suitable categories for an appropriate description of God's time.

Christian hope presupposes both what modern science recognizes more and more clearly and what has a high degree of reasonableness philosophically:[2] As a system open to the future, time is constituted and moves from this openness.

---

1. The process of the reception of chaos theory by theology has already begun (Ganoczy 1995; Russell, Murphy, and Peacocke 1995; Achtner 1997).

2. Cf. the astute article by Karl R. Popper (1993, 172-82).

THE TRI-POLAR STRUCTURE OF TIME AND LOVE

As we saw, the biblical concept of time differs from the mythic, the rational-linear, and the mystic concepts in that it balances time in its tri-polar structure. The endogenous, exogenous, and transcendent experiences of time caused and corrected one another. This result stands in a surprising proximity to Jesus' command to love, which is the basis of all Christian ethics. It is known as the "double commandment of love," but with more exactness it is stretched into three commandments: "You shall love the Lord your God with all your heart, and with all your soul, and with all your strength, and with all your mind; and your neighbor as yourself" (Luke 10:27 [NRSV]) — that is, God, neighbor, and self.

The command to love God implies the openness of God's own infinite love and the readiness of God to "be loved" most deeply. This experience is the basis of a healthy love of self, and the love of self is the basis of a helpful love of the neighbor. In terms of our model: as the experience of an "unconditional" love approaching me, the experience of transcendent time makes the experience of endogenous time possible, in the sense of an experience of being loved in the rhythm of my own individual existence. Yet this experience also enables the experience of exogenous time as loving attention to the temporal reality in nature and history. The tri-polar structure of time and the tri-polar commandment of love by Jesus mutually expound each other. "Being" and "shall be" correspond to each other in the Spirit of Jesus Christ in a tri-polar analogy.

The Experience of a "Fulfilled Time"

OUTLOOK

It was the goal of our investigation

- to make comprehensible the structure of bygone human experiences of time,
- to arrange meaningfully the variety of present human experiences of time, and
- to show ways of solving the temporal crisis of our "accelerated society."

The first part ("The Time of Humans") showed in terms of anthropology and intellectual history that the structure of human experiences of time in the past could be deciphered by using the model of tri-polar time: Synchronism and asynchronism in the tri-polar (as well as, in the deficit situation, bipolar) structure of time make comprehensible how people experienced time in the course of the history of culture. It shows how they set time in motion, and how they dealt with it. Stability and changes in cultures by it were reasonable, in a certain sense.

It became clear that the forms of the perception of endogenous time developed toward exogenous and transcendent time by synchronism and asynchronism. They could be represented in the form of an "evolution of consciousness."

In the second part ("The Time of the World"), which was concerned with physics, it became clear that for its concept of time modern physics assumed different aspects of the three stages of the experience of endogenous time. Modern physics, as it were, verified it in nature. Thus, certain analogies exist between the experience of mythic-cyclic time and the symmetry of time in most physical laws. The experience of rational-linear time can be compared with the conception of time as a linear parameter, and likewise as the symmetry of time. The holistic experience of time corresponds to the connection of space and time in a four-dimensional space-time, in the theory of relativity and modern cosmology.

However, it turned out that L. Boltzmann burst all three forms of the experience of endogenous time with the arrangement of time in thermodynamics. The discovery of irreversibility in nature forces a revision of all traditional concepts of time. Time has an irreversible direction, and, according to the structure of the system to which it belongs, it is suited to an ever richer and more open structure! Here we come into a close proximity to the understanding of time and history in the Jewish-Christian tradition.

In a third part dealing with theology ("The Time of God"), it was shown how time was experienced and understood in the central traditions of the Old and New Testaments. It is an open, tri-polar system.

We have laid the basis for classifying and interpreting the current human experience of time in our Western, technological societies. If we apply our tri-polar system to our present society, then endogenous time is seen to be quite disproportionately dominated by rational-linear time. Transcendent time plays such a small role that exogenous time is reformed and outwitted by the omnipotence of the linear time of the clock. The cycles of nature must bow to the dictation of linear time. Even the cycles of the growth of plants must be fitted into this inherent corset of time (strawberries in the winter, etc.). To put it briefly, our tri-polar system has been thrown completely out of balance. If we subject our present concept of linear time to a more exact analysis, then some phenomena of modern times could be diagnosed as a kind of illness of time:

- Linear time is a quantifiable continuum that is divisible at will (atomic clocks). Since it is quantifiable, it is a commodity with a corresponding economic value ("time is money"). In the language of economics, time spent in production causes cost, and that involves the pressure of rationalization. This phenomenon is frequently noticed in the acceleration of life in modern times. As Fritz Reheis (1995) has presented in detail, we live in an "accelerated society." In always-faster cycles it strives for an increase in production and consumption. This condition causes more and more illnesses in people, in both mind and

174

body. It causes social relationships to disintegrate and natural resources to dry up. The "acceleration society" seems to evoke a "syn-
drome of hyper-hype": Not only individuals but also the whole society, and even our biosphere, are permanently overtaxed by an
accelerated exogenous time. They are aggressively or regressively
driven to react to the pressure of hyper-hype.

- The effect of acceleration leads to a complementary effect of things
becoming obsolete. Hermann Lübbe (1989, 140-49) has impressively
explained in greater detail how the waves of nostalgia reach closer and
closer to present actuality. One could exaggerate ironically by saying
that the most nostalgia-filled people today live in the future in the
present. This effect is true for the cycles of innovation in fashion, science, slogans, the so-called public debates that rarely lead to a tangible
result, styles of music, the (un)culture of clothing, hairstyles, and so
forth. For various waves of fashion, the effect of rapid obsolescence
may continue. In economic affairs, it is a very serious problem, however. The time for the amortization of investments is ever shorter, because of the pressure for innovation. In turn, that provides worry
about a new acceleration.

- Linear, quantifiable time is free of qualities. Biblical humans still had
an inner feel for the right time ("For everything there is a season . . . ,"
Ecclesiastes 3), just as the ancient Greeks had for *kairos*. It appears that
this inner perception of the quality of time for such actions, or such
inner events, has been neglected in favor of the quantifiable time of rational plans. The elementary feeling for age-appropriate behavior is
endangered, as well as the knowledge that human life itself happens in
biological, spiritual, and intellectual cycles of maturity. Furthermore,
the strengths that form personality are not perceived. When quantifying time hinders the interior guidance of humans, their interior qualities waste away. Accordingly, in place of the qualitative, temporal, and
interior guidance is found the quantitatively outer guidance (higher,
faster, and further) of humans. The average human is guided more by
the external social clocks in the form of consumption, cycles of work,
cycles of vacation, cycles of television, and cycles of fashion: *cycles!*

- In the somewhat complicated detour of the endangerment of the
qualitative, inner guidance[1] by quantitative, linear time, there is a re-

1. This loss of the inner temporal guidance and the dependence on external social

175

activation of the structures of mythic experience. It is not rare to find counterparts — from the "myth of the twentieth century," to modern fantasy novels and the glorification of mythic cultures in the Celtic and Native American renaissance (at the moment, the ancient German cult is probably only in fringe groups). This external guidance pushes humans into a dangerous passivity and dependence on external social clocks. Those situations easily turn into the addiction for increasingly faster and larger doses of satisfaction. So, modern humans are caught in the prison of this continuous repetition of secularized mythic structures, which keep them away from a real confrontation with their environment. This provides very little for the ability for maturing the self, as well as for the acquisition of a certain sovereignty over time. The latter is necessary to build up a tolerance for frustration in order to abstain from instant gratification in favor of a higher goal. The delay of the satisfaction of needs is no longer a virtue in our list of values. If wishes must be instantly gratified, best of all "online," then disappointments will quickly turn into aggression or frustration.

- This qualified, external, and linear time overlaps the qualitative, inward, biological rhythms. We have already encountered this phenomenon with shift work and jet lag. The superimposition of suitable experience is possible only within certain tolerance values; otherwise, it becomes unhealthy. Also, the pressures of a deadline, and the pressures of making a decision in a quantified time, have consequences for health (illness of managers, psychosomatic disturbances). For the management of business, it would be worthwhile to compare the cost advantage of time-saving with the damage economically in work-related illnesses. However, we see the most serious disadvantage in its not being conducive to the creative processes in the unconscious depths of humans, who have their own dynamics of time.

Ways out of this "accelerated society" and its consequent phenomena are discussed everywhere, and approaches are being tried. On the basis of our investigations, we can divide them into three basic forms:

---

guidance are shown in joblessness. Investigations prove that jobless people lose the ability to form their time meaningfully (cf. Schräder-Naef 1989, 17-25).

- Ways out by mythic regression,
- Ways out by rational-linear technology, and
- Ways out by mystic-holistic escapism.

The first way out appears, above all, in the return to supposedly "natural" forms of life. It revives the ancient pre-Renaissance lifestyle of esoteric religiosity, in a fight against science and technology, and alternative endeavors at autarchy and irrational and magical methods.

The second way out is a technocratic imagination of omnipotence. It hopes to get a grip on the problems of the "accelerated society" by overcoming the risks by means of technology. With their belief in progress, these representatives see a way out of the crisis by a "strong artificial intelligence."

The third way out tries by mystic transcendence and present experience to be set free from the hyper-hype of "accelerated time." The increasing interest in meditation, contemplation, and Eastern philosophy and religion shows how very much these ways are sought and practiced.

However, our own approach allows us critically to question these ways out of the "accelerated society":

Mythic regression seems to want to understand time as a closed system — contrary to the realizations of modern physics and the immediate experience of time — which would give it protection and security by its symmetry and synchronism. This attempt seems to be possible only in view of an enormous loss of reality in exogenous time. It risks being burned out in the face of renewing transcendent time.

The *rational-linear technocracy*, however, misjudges time with the fullness of the interlacing of its modes. It still proceeds from a one-dimensional, linear concept of time. It disregards the "characteristic times" of the human experience of time in the endogenous area, such as the unpredictability and unavailability of transcendent time as a source of new life.

Mystic escapism has finally reduced the tri-polar structure of time to a synchronization of the transcendent and endogenous experience of time. It happens because of a complete contempt for the exogenous experience of time in nature and history. The balance in the tri-polar structure of time is no longer maintained, and the structure shrivels to a single point of the mystic experience of eternity in the "here and now."

For these previous ways out, we would not like to propose any alternative solutions as recipes for pre-cooked meals. However, we do encourage a confrontation, especially with modern science and the Jewish-Christian tradition. This confrontation would advance the discussion and bring the possibility of successful solutions into the social discourse. There are at least three minimal requirements that we defined by our investigations. These requirements might yield a framework, which in our opinion would make possible and meaningful an adequate discussion of time. The three minimal requirements are:

- Time must (once again) be perceived, conceived, and experienced as an *open system*.
- Time must (once again) be perceived, conceived, and experienced in the fullness of *its modes and in the interlacing of its modes*.
- Time must (once again) be perceived, recognized, and balanced in its tri-polar structure.

Despite the dubious dominance of linear time, we are not demonizing the concept of linear time. On the contrary, we consider it to be a wonderful intellectual achievement of Western thought. It is possible to integrate it appropriately into the threefold human experience of endogenous time, and into the entire tri-polar system. Concretely, it means to open downward into the emotional depths of mythic-cyclic time, as well as upward into mystic-holistic time. However, this opening to mythic-cyclic time is not required for the regressive tendencies, which were described earlier. Rather, it is a controlled, consciously formed "regression in the service of the self." It grants me an emotional feeding, it supplies the regenerative and creative strength of my *own time*, and it protects me from intellectual sterility. That happens only through a conscious contact with the unconscious, for example, with what Sigmund Freud suggested was "consistently undecided attention," when the self releases itself from the limits of linear time. What would such a withdrawal from linear time look like? Into what form of time can the self change itself? These things may be made clear in the following from Roberto Juarroz:

Thirteenth Vertical Poetry (52)

Today, I did nothing.
But many things happened in me.

A bird, who didn't have one,
Found a nest.

Shadows that might be there,
They gained bodies.
Words that exist
Regained their peace.

To do nothing,
sometimes rescues the balance of the world,
in that it reaches something important
on the empty shell of the scales.

At their leisure, readers may interpret this poem in terms of our tri-polar system. In any case, it presents a creative contact with the course of time, when the external guidance of linear time is set aside and the inner guidance is at work. Such an inner guidance seems to function by integrating unconsciously, and in one's own life, the disorderly endogenous "time of the self," but it is a source of a creative lifestyle. Historically, many of the great intellectual achievements were built, so to speak, on such islands of time. Isaac Newton developed his mechanics in rural isolation in his hometown of Wollsthorpe, while the plague raged in London. As for giving him the idea for developing the law of gravity, the story of the falling apple was just an invention; yet in any case he had leisure. Albert Einstein also had leisure in his patent office, far away from the great business of science. In quiet confrontation with the classical concept of time, he discovered in the truest sense of the word his "own time." The most recent transformation of the concept of time has occurred in the context of the fuzziness of quantum mechanics. It was a fortunate idea of the young Werner Heisenberg during a time-out on the island of Helgoland (1973, 76-77). For all creative processes, such time-outs seem to be necessary for creating a leap from one view to another.[2]

How much would be gained for our society if such time-outs were not feared as an *end*, but rather they were tolerated and accepted as chances for new beginnings. Instead, those who have time are looked at compassionately as "out of it." It would be good not only to accept leisure as forming social time, but also to use its regenerative and creative potential, in the sense of regressing in the service of the self. Leisure need not be called idleness.

2. There are investigations in different sciences: for example, for mathematics (Hadamard 1954), for mathematics and physics (Wertheim 1964), and for general creativity (Landau 1984).

Also, broadly in the history of culture, it is good that a retreat in time sets free creative power, and it can show new ways. What would have happened in antiquity without the power of retreat by Benedict of Nursia? How would the West have developed without the life of "pray and work!"? How would the treasures of Western science ever have been able to develop without the cultivation of peace, and without the diligence of monastic learning in the medieval cloisters? At least in these places, space was given for the mental purification of endogenous time, and for opening new possibilities of mystic-holistic and transcendent time.

Briefly, the balanced interlacing of linear time into cyclic-mythic time (as a regression in the service of the self) and into the mystic-holistic time (as a progression by the opening of the self) is practiced anew in the face of transcendent time. This practice can absolutely happen in everyday life, in accordance with the journey of a circular process of new birth and return, and it can also happen in professional life. It is not a question of having no time, but it is a question of priorities.

This "balance" in the tri-polar structure of time seems to us to be necessary for survival. It implies the "deceleration" of life by a return to the "own time," which is claimed for all human lives (in a more elastic manner). It implies the synchronization of historical time and natural time, of economic time and ecological time, as a long-term objective. It implies openness in the face of "God's time" as "the source of life," as it is experienced in faith.

The goal is the experience of "fulfilled time," which should be the most precious thing that can be given to us.

BIBLIOGRAPHY

Achtner, W. 1991. *Physik, Mystik und Christentum: Eine Darstellung und Diskussion der natürlichen Theologie bei T. F. Torrance.* Frankfurt.

————. 1997. *Die Chaostheorie — Geschichte — Gestalt — Rezeption.* EZW — Texte 135. Berlin.

————. 2001. "Zeit und Ewigkeit aus Religiöser und aus Christlicher Sicht." In *Zeit und Eigenzeit als Dimensionen der Sonderpädagogik.* Ed. C. Hofmann et al., pp. 379-89. Luzern.

Albrecht, C. 1951. *Psychologie des mystischen Bewußtseins.* Bremen.

————. 1968. *Akhenaten King of Egypt.* London.

Aldred, C. 1968. *Akhenaten King of Egypt.* London.

Alt, A. 1970. "Der Gott der Väter." In *Grundfragen der Geschichte des Volkes Israel,* pp. 21-98. Munich.

Aschoff, J., and J. Wever, eds. 1981. *Handbook of Behavioral Neurobiology,* vol. 4. New York.

Assmann, J. 1975. *Zeit und Ewigkeit im alten Ägypten.* Heidelberg.

————. 1991. *Stein und Zeit, Mensch und Gesellschaft im alten Ägypten.* Munich.

————. 1996. *Ägypten, Eine Sinngeschichte.* Munich.

Banks, J., and J. Brooks, G. Cairns, G. Davis, and P. Stacy. 1992. "On Devaney's Definition of Chaos." *The American Mathematical Monthly* 99 (4).

Barr, James. 1962. *Biblical Words for Time.* London. Rev. ed., 1969.

Beckerath, E., ed. 1956. *Handbuch der Sozialwissenschaften.* Göttingen.

Bettini, O. 1953. "La temporalita della cosa e l'esigenza di un principio assoluto nella dottrina di Olivi." In *Antonianum.* Rome.

Beyerlin, W., ed. 1975. *Religionsgeschichtliches Textbuch zum Alten Testament.* Göttingen.

Bilfinger, G. 1969. *Mittelalterliche Horen und die modernen Stunden: Ein Beitrag zur Kulturgeschichte.* Wiesbaden.

181

Blume, F., ed. 1989. *Musik in Geschichte und Gegenwart (MGG)*. Kassel.

Blumenberg, H. 1974. *Säkularisierung und Selbstbehauptung: Die Legitimität der Neuzeit*. New ed. Frankfurt.

———. 1981. *Die Genesis der kopernikanischen Welt*. Band 1-3. Frankfurt.

Boltzmann, L. 1896. *Vorlesung über Gastheorie*. Leipzig.

Burger, H. 1986. *Zeit, Natur und Mensch*. Berlin.

Capra, F. 1980. *Der kosmische Reigen*. Munich.

Coleman, R. M. 1986. *Wide Awake at 3:00 A.M., by Choice or by Chance*. New York.

d'Aquili, Eugene G., and Andrew B. Newberg. 1999. *The Mystical Mind: Probing the Biology of Religious Experience*. Minneapolis.

Devaney, R. L. 1989. *An Introduction to Chaotic Dynamical Systems*. 2nd ed. Redwood City, Calif.

Dohrn-van Rossum, G. 1989. "Schlaguhr und Zeitorganisation." In *Im Netz der Zeit*. Ed. R. Wendorff, pp. 49-60. Stuttgart.

Eggebrecht, H. H. 1991. *Musik im Abendland*. Munich.

Einstein, A. 1922. *Grundzüge der Relativitätstheorie*. Braunschweig.

Engel, K. 1995. *Meditation: Geschichte — Systematik — Forschung — Theorie*. Frankfurt.

Fick, E. 1988. *Einführung in die Grundlagen der Quantenmechanik*. 6th ed. Wiesbaden.

Fränkel, H. 1931. *Die Zeitauffassung in der frühgriechischen Literatur*. Leipzig.

Frankfort, H. 1955. *Kingship and the Gods*. Chicago.

Frankfort, H. A., J. A. Wilson, T. Jacobsen, and W. Irwin. 1954. *Frühlicht des Geistes*. Stuttgart.

Fraser, J. T. 1972. *The Study of Time*. Berlin, Heidelberg, and New York.

Ganoczy, A. 1995. *Chaos, Zufall, Schöpfungsglaube: Die Chaostheorie als Herausforderung der Theologie*. Mainz.

Gerbert, M. 1963. *Scriptores Ecclesiastici de Musica sacra potissimum*. Band 3. Hildesheim.

Gericke, H. 1993. *Mathematik in Antike und Orient*. Wiesbaden.

Gillette, M. U. 1991. "SCN Electrophysiology in Vitro: Rhythmic Activity and Endogenous Clock Properties." In *Suprachiasmatic Nucleus: The Mind's Clock*. Ed. D. C. Klein and R. Y. Moore. New York and Oxford.

Glasser, R. 1936. *Studien zur Geschichte des französischen Zeitbegriffs*. Munich.

Goethe, J. W. v. 1994. *Faust*. Frankfurter Ausgabe Band 7. Frankfurt.

Goldstein, H. 1985. *Klassische Mechanik*. 8th ed. Wiesbaden.

Gronemeyer, M. 1993. *Das Leben als letzte Gelegenheit*. Darmstadt.

Grüsser, O.-J. 1992. "Zeit und Gehirn, Zeitliche Aspekte der Signalverarbeitung in den Sinnesorganen und im Zentralnervensystem." In *Die Zeit, Dauer und Augenblick*. Ed. H. Gumin and H. Meier, pp. 79-132. Munich.

Guckenheimer, J., and P. Holmes. 1986. *Nonlinear Oscillations, Dynamical Systems, and Bifurcations of Vector Fields*. 2nd ed. New York.

Gumin, H., and H. Meier, eds. 1989. *Die Zeit, Dauer und Augenblick*. Munich.

Bibliography

Hadamard, J. 1954. *The Psychology of Invention in the Mathematical Field.* Princeton, N.J.

Haranjo, C., and B. E. Ornstein. 1988. *Psychologie der Meditation.* Frankfurt.

Heimann, H. 1992. "Zeitstrukturen in der Psychopathologie." In *Die Zeit, Dauer und Augenblick.* Ed. H. Gumin and H. Meier, pp. 59-78. Munich.

Heisenberg, W. 1973. *Der Teil und das Ganze.* Munich.

Huizinga, J. 1957. *Herbst des Mittelalters.* Studien über Lebens (und Geistesformen des 14. und 15.Jahrhunderts in Frankreich und in den Niederlanden). Stuttgart.

Hund, F. 1984. *Geschichte der Quantentheorie.* Mannheim.

Huxley, A. 1986. *Die Pforten der Wahrnehmung, Himmel und Hölle, Erfahrungen mit Drogen.* Munich.

Illnerova, H. 1991. "The Suprachiasmatic Nucleus and Rhythmic Pineal Melatonin Production." In *Suprachiasmatic Nucleus: The Mind's Clock.* Ed. D. C. Klein and R. Y. Moore. New York and Oxford.

James, William. 1958. *The Varieties of Religious Experience.* New York.

Jaspers, K. 1964. *Nikolaus Cusanus.* Munich.

Jastrow, M., Jr. 1905. *Die Religion Babylons und Assyriens.* Band 2. Gießen.

Jauch, J. M., and J. G. Baron. 1961. "The Many Faces of Thermodynamics and Statistical Physics." *Communications in Pure and Applied Science* 14: 323ff.

Jenni, E., and C. Westermann. 1979. *Theologisches Handwörterbuch zum Alten Testament.* Munich.

Jüngel, E. 1972. *Unterwegs zur Sache: Theologische Bemerkungen.* Munich.

Kant, I. 1990. *Kritik der reinen Vernunft.* Hamburg.

Kapleau, P. 1989. *Die drei Pfeiler des Zen.* Munich.

Kirshner, J., and K. L. Prete. 1984. "Peter John Olivi's Treatises on Contracts of Sale Usury, and Restitution: Minorite Economics or Minor Works?" *Quaderni fiorentini* 13.

Koselleck, R. 1973. "Geschichte, Geschichten und formale Zeitstrukturen." In *Geschichte: Ereignis und Erzählung.* Ed. R. Koselleck and W. D. Stempel. Munich.

Kroll, G. 1988. *Auf den Spuren Jesu.* Leipzig.

Kunz, S. 1985. *Zeit und Ewigkeit bei Meister Eckhart.* Tübingen.

Landau, E. 1984. *Kreatives Erleben.* Munich.

Laugwitz, D. 1986. *Zahlen und Kontinuum.* Mannheim.

Lebram, J. 1978. "Apokalyptik." In *Theologische Real-Enzyklopädie (TRE).* Band 3, pp. 192-202. Berlin.

Lemmer, B. 1983. *Chronopharmakologie.* Stuttgart.

Lichtenberg, A. J., and M. A. Lieberman. 1983. *Regular and Stochastic Motion.* New York.

Liedke, G. 1974. *Zeit, Wirklichkeit und Gott, Referat, gehalten am 27.2.1974 im Studium Generale der Universität Utrecht.* Utrecht.

Link, C. 1978. *Subjektivität und Wahrheit.* Stuttgart.

Lübbe, H. 1989. "Zeit-Verhältnisse: Über die veränderte Gegenwart von Zukunft und Vergangenheit." In *Im Netz der Zeit.* Ed. R. Wendorff, pp. 140-49. Stuttgart.

Luhmann, N. 1975. *Weltzeit und Systemgeschichte*. Soziologische Aufklärung 2. Opladen.

Mahnke, R., J. Schmelzer, and G. Röpke. 1992. *Nichtlineare Phänomene und Selbstorganisation*. Stuttgart.

Maier, A. 1949. *Die Vorläufer Galileis im 14.Jahrhundert: Studien zur Naturphilosophie der Spätscholastik*. Storia e Litteratura 22. Rome.

———. 1955. *Metaphysische Hintergründe der spätscholastischen Naturphilosophie*. Storia e Litteratura 52. Rome.

Mann, U. 1985. *Schöpfungsmythen*. Stuttgart.

Maslow, A. 1989. *Psychologie des Seins*. Frankfurt.

Meier-Koll, A. 1995. *Chronobiologie*. Munich.

Mittelstaedt, P. 1963. *Philosophische Probleme der modernen Physik*. Mannheim.

———. 1989. *Der Zeitbegriff in der Physik*. Mannheim.

Mletzko, H. G., and I. Mletzko. 1985. *Biorhythmik*. Wittenberg.

Moltmann, J. 1964. *Theologie der Hoffnung*. Munich.

———. 1985. *Gott in der Schöpfung: Ökologische Schöpfungslehre*. Munich.

———. 1995. *Das Kommen Gottes: Christliche Eschatologie*. Gütersloh.

Müller, A. M. K. 1972. *Die präparierte Zeit*. Stuttgart.

Nestler, G. 1975. *Geschichte der Musik*. Gütersloh.

New Revised Standard Version Bible (NRSV). 1989. Nashville, Tenn.

Newton, I. 1952. *Mathematical Principles of Natural Philosophy, Optics*. Trans. A. Motte, rev. F. Cajori. Vol. 34 of *Great Books of the Western World*, ed. R. M. Hutchins. Chicago.

———. 1988. *Mathematische Grundlagen der Naturphilosophie*. Hamburg.

Nietzsche, F. 1872. *Die Geburt der Tragödie aus dem Geist der Musik*. Stuttgart.

North, J. D. 1976. *Richard of Wellington*. Oxford.

Noth, M. 1950. *Geschichte Israels*. Göttingen.

Ockham, G. de. 1987. *Brevis summa libri physicorum (Summula Philosophiae Naturalis et Questiones in libros physicorum Aristotelis)*. Ed. S. Brown. New York.

Oeser, E., and F. Seitelberger. 1988. *Gehirn, Bewußtsein und Erkenntnis*. Darmstadt.

Olivi, P. J. 1980. *Petrus Johannis: Un trattato di economia politica francescana: il "De emptionibus et venditionibus, de usuris, de restitutionibus" di Pietro di Giovanni Olivi*. Ed. Giacomo Todeschini. Rome.

———. 1990. *Petrus Johannis: Usure, compere e vendite: la scienca economica des XIII secolo*. Pietro di Giovanni Olivi: A cura di Amleto Spicciani. Milan.

Oresme, N. 1503. *Tractatus de origine, natura iure et mutationibus monetarum (Neuausgabe lateinisch und deutsch von E. Schorer)*. Paris.

Origo, I. 1989. *Der Heilige der Toskana, Leben und Zeit des Bernardino von Siena*. Munich.

Perler, D. 1988. *Prädestination, Zeit und Kontingenz*. Amsterdam.

Piaget, J. 1955. *Die Bildung des Zeitbegriffs beim Kinde*. Zürich.

Picht, G. 1980. "Die Zeit und die Modalitäten." In his *Hier und Jetzt: Philosophieren nach Auschwitz und Hiroshima*. Band 1, pp. 362-75. Stuttgart.

Bibliography

Plutarch. 1777. *De Iside et Osiride.* Lipsiae.

Pöppel, E. 1985. *Grenzen des Bewußtseins.* Stuttgart.

Popper, K. R. 1993. "Ludwig Boltzmann und die Richtung des Zeitablaufs: Der Pfeil der Zeit." In *Klassiker der modernen Zeitphilosophie.* Ed. W. Zimmerli and M. Sandbothe, pp. 172-81. Darmstadt.

Prigogine, I. 1970. "Dynamical Foundations of Thermodynamics and Statistical Mechanics." In *A Critical Review of Thermodynamics and Statistical Physics.* Ed. B. Gal-Or and A. J. Brainard. Baltimore.

————. 1995. *Die Gesetze des Chaos.* Frankfurt.

Prigogine, I., and I. Stengers. 1992. *Vom Sein zum Werden.* Munich.

Prigogine, I., and I. Stengers. 1993. *Das Paradox der Zeit.* Munich.

Prior, A. 1967. *Past, Present and Future.* Oxford.

Quint, J. 1963. *Meister Eckhart.* Munich.

Rad, G. 1980. *Theologie des Alten Testaments.* Band 2. Munich.

Reheis, F. 1995. *Die Kreativität der Langsamkeit.* Darmstadt.

Rinderspacher, J. P. 1989. "Mit der Zeit arbeiten." In *Im Netz der Zeit.* Ed. R. Wendorff, pp. 91-104. Stuttgart.

Rindler, W. 1977. *Essential Relativity (Special, General, and Cosmological).* New York.

Robinson, G. 1988. *The Origin and Development of the Old Testament Sabbath.* Frankfurt.

Roover, R., and J. A. Schumpeter. 1957. "Scholastic Economics." In *Kyklos* 2.

Roth, G. 1995. *Das Gehirn und seine Wirklichkeit.* Frankfurt.

Rubber, B. 1948. *Über sogenannte kosmische Rhythmen beim Menschen.* Stuttgart.

Russell, R. J., N. Murphy, and A. R. Peacocke. 1995. *Chaos and Complexity.* Vatican City State.

Sarton, G. 1953. *The History of Science.* Vol. 2. Baltimore.

Schleiermacher, F. 1924. *Über die Religion: Reden an die Gebildeten unter ihren Verächtern.* Leipzig.

————. 1994. *On Religion: Speeches to Its Cultured Despisers.* Trans. J. Oman. Louisville, Ky.

Schmidt, W. H. 1988. *Biblischer Kommentar.* Band 2/1, 1. Teilband. Neukircher.

Schnabel, P. 1925. "Der jüngste Keilschrifttext." *Zeitschrift für Assyriologie* n.s. 2 (36): 66ff.

Schneider, I. 1988. *Isaac Newton.* Munich.

Schott, A., ed. 1980. *Das Gilgameschepos.* Stuttgart.

Schräder-Naef, R. 1989. "Zeit als Belastung?" In *Im Netz der Zeit.* Ed. R. Wendorff, pp. 17-25. Stuttgart.

Schüttler, G. 1968. *Das Mystisch-Ekstatische Erlebnis: Systematische Darstellung der Phänomenologie und des Psychopathologischen Aufbaus.* Bonn.

————. 1974. *Die Erleuchtung im Zen-Buddhismus.* Munich.

Seleschnikow, S. I. 1981. *Wieviel Monde hat ein Jahr?* Moskau.

Sexl, R. U., and H. K. Urbantke. 1987. *Gravitation und Kosmologie.* Zürich.

185

Shapiro, H. 1957. *Motion, Time and Place According to William Ockham, St. Bonaventure.* New York.

Stadler, M., P. Kruse, and H.-O. Carmesin. 1996. "Erleben und Verhalten in der Polarität von Chaos und Ordnung." In *Chaos und Ordnung, Formen der Selbstorganisation in Natur und Gesellschaft.* Ed. G. Küppers, pp. 323-52. Stuttgart.

Storch, A. 1922. *Das archaisch-primitive Erleben und Denken der Schizophrenen.* Berlin.

Theißen, G. 1994. *Lichtspuren, Predigten und Bibelarbeiten.* Gütersloh.

Thomas, K. 1973. *Meditation in Forschung und Erfahrung.* Stuttgart.

Tölle, R. 1991. *Psychiatrie.* Heidelberg.

Vaux, R. de. 1960. *Das Alte Testament und seine Lebensordnungen.* Vol. 2. Freiburg.

Walter, T. 1994. *Verzögerte Bifurkationen als Effekt langsamer Parameteränderung.* Aachen.

Walter, T., F. Rödelsperger, and H. Benner. 1996. "Delayed Bifurcations at the First-order Suhl Threshold." *Zeitschrift für angewandte Mathematik und Physik (ZAMP)* 47: 515-26.

Weber, M. 1988. "Die protestantische Ethik und der Geist des Protestantismus." In *Gesammelte Aufsatze zur Religionssoziologie I,* pp. 17-206. Tübingen.

Weizsäcker, C. F. v. 1971. *Die Einheit der Natur.* Munich.

Wendorff, R. 1985. *Zeit und Kultur: Geschichte des Zeitbewußtseins in Europa.* Opladen.

Wertheim, M. 1964. *Produktives Denken.* London.

Westermann, C. 1981. *Genesis,* vol. 2. Biblische Kommentar Altes Testament. Band 1/2. Teilband. Neukirchen.

Whitrow, G. J. 1973. *Von nun an bis in Ewigkeit.* Düsseldorf.

————. 1991. *Die Erfindung der Zeit.* Hamburg.

Wiggins, S. 1990. *Introduction to Applied Nonlinear Dynamical Systems and Chaos.* New York.

Wissowa, G., and W. Kroll, eds. 1930. *Real-Encyclopädie der classischen Altertumswissenschaft.* Stuttgart.

Wolff, H. W. 1969. *Dodekapropheton,* vol. 2: *Joel und Amos.* Biblischer Kommentar Altes Testament Band 14/2. Neukirchen.

————. 1973. *Anthropologie des Alten Testaments.* Munich.

Zeh, H. D. 1984. *Die Physik der Zeitrichtung.* Vol. 200 of *Lecture Notes in Physics.* Heidelberg.

Zucker, F. J. 1974. "Information, Entropie, Komplementarität und Zeit." In *Offene Systeme,* vol. 1. Ed. E. u. von Weizsäcker, pp. 35-81. Stuttgart.

INDEX OF SUBJECTS

Apocalypticism, 148-51, 149n, 165; and endogenous time, 150, 151; and exogenous natural time, 150-51; and linear time, 150; and transcendent time, 149-51

Astrology: and apocalypticism, 150, 151; and Mesopotamia, 40-45, 50, 53

Biological bases of experience of time, 12-26; and biological rhythms, 12-18, 176; and chaos theory, 16, 135, 136; and consciousness, 18-26; and coupling of endogenous and exogenous time, 16-18; endogenous time and, 12-18, 24, 25-26; exogenous time and, 12, 16-18; and mystic-holistic time, 24, 25-26; and mythic-cyclic time, 24-26; and rational-linear time, 24, 25-26, 176; and suprachiasmatic nucleus, 16-18; and transcendent time, 26

Biological rhythms, 12-18, 176; and coupling of endogenous and exogenous time, 16-18; endogenous, 12-18; exogenous, 12, 13-15; and linear time in present-day society, 176; and psychic illnesses, 14-16; and suprachiasmatic nucleus, 16-18; and synchronism and asynchronism, 14-16

Calendars: and Greek measurement of time, 65-66; Mesopotamian lunar, 41-43

Chaos theory and time, 128-36, 137; definitions of chaos, 128-31; and delayed bifurcations, 133; disturbances and dynamic systems, 135-36; and human biological systems, 16, 135, 136; the KAM Theorem, 134-35; and meditation, 25n; and open systems, 171; Poincaré's Picture I, 131; Poincaré's Picture II, 132-33

Clocks: disconnecting time from natural rhythms, 85-86; and Greek measurement of time, 65-66; invention of, 65, 76-77, 82-83, 84-86, 91; and the plague, 91; in present-day society, 174; and relationship between God and time, 86

Consciousness and time, 18-26; and biological bases of time, 18-26; and Egypt, 32-33, 37-39; endogenous time, 8-9, 8n, 24-26, 30-35, 173; evolution of consciousness of time, 8-9, 8n, 19-21, 24-26, 30-35, 37-39, 40, 57, 173; and Hebrew nomads, 57, 61; and individualization, 19;

191